Environmental Remote Sensing in Flooding Areas

Chunxiang Cao · Min Xu · Patcharin Kamsing ·
Sornkitja Boonprong · Peera Yomwan ·
Apitach Saokarn

Environmental Remote Sensing in Flooding Areas

A Case Study of Ayutthaya, Thailand

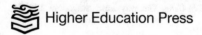 Higher Education Press

 Springer

Chunxiang Cao
Aerospace Information Research Institute
Chinese Academy of Sciences
Beijing, China

Patcharin Kamsing
International Academy
of Aviation Industry
King Mongkut's Institute
of Technology Ladkrabang
Bangkok, Thailand

Peera Yomwan
Department of Lands
Ministry of Interior
Bangkok, Thailand

Min Xu
Aerospace Information Research Institute
Chinese Academy of Sciences
Beijing, China

Sornkitja Boonprong
Department of Geography
Faculty of Social Sciences
Kasetsart University
Bangkok, Thailand

Apitach Saokarn
Royal Thai Survey Department
Bangkok, Thailand

ISBN 978-981-15-8204-2   ISBN 978-981-15-8202-8   (eBook)
https://doi.org/10.1007/978-981-15-8202-8

Jointly published with Higher Education Press
The print edition is not for sale in China (Mainland). Customers from China (Mainland) please order the print book from: Higher Education Press.
ISBN of the Co-Publisher's edition: 978-7-04-055317-8

This Springer imprint is published by the registered company Springer Nature Singapore Pte Ltd.
The registered company address is: 152 Beach Road, #21-01/04 Gateway East, Singapore 189721, Singapore

Foreword

Floods are among the most devastating natural hazards in the world, far and wide leading to more loss of life, significant economic and social damages than any other natural phenomenon. Under the influence of climate change and economic development, the extent of flooding is expected to increase. The primary effects of flooding include loss of life and damage to buildings and other structures, including bridges, sewer systems, roadways, and canals. Infrastructure damage also frequently damages power transmission and at times power generation, which then has resultant effects caused by the loss of power. This includes loss of drinking water treatment and water supply, which may result in loss of drinking water or severe water contamination. Flooding is associated with an increased risk of infection. However, this risk is low unless there are significant population displacement and/or water sources compromised. The risk of infection from waterborne diseases may increase substantially during periods of flooding. The risk of disease has typically been assessed by the analysis of pathogens in samples of contaminated floodwater. Waterborne disease outbreaks have been associated with periods of heightened source water pathogen concentrations in treated drinking water supplies. For their management it is necessary to identify and quantify the impacts of events which lead to adverse concentration fluctuations.

The World Academy of Sciences (TWAS) has collaborated with the Chinese Academy of Sciences (CAS) to conduct research on mitigation of the effects of flood disaster under joint research projects, such as Asian Space Technologies for Disaster Mitigation by CAS-TWAS Center of Excellence on Space Technology for Disaster Mitigation (SDIM). Moreover, the aim to reduce flooding effects moved the governments of several countries, such as Thailand and China, to support related research. Central Thailand, especially Chao Phraya basin, has been frequently affected by flooding during the monsoon season, while the Yangtze River basin of China also experiences the same problem. Therefore, joint research on flooding was raised. The important thing after a flood is flood loss estimation to provide proper help to people in need. This requires precise knowledge of flooding areas in order to contribute to such help. Thus, flood classification plays a crucial role in this aspect.

This book contains the experimental results from the study of flood identification to accurately estimate flooding, which would be beneficial for flood loss estimation and other applications. The waterborne disease caused by flooding also illustrates the results in the second part of this book. It also provides reference value for the development of further research in flooding.

July 2020   Prof. Ramesh P. Singh
                                                                                    Chapman University
                                                                                          Orange, USA

Preface

This book contains a collection of research works in regard to the major flood in Thailand in 2011. Ayutthaya Province in central Thailand is an area of study and an important province for both the economy and many archaeological sites of Thailand. Several projects have investigated the cause of the disaster and finding the method to mitigate the impact. The results illustrated in this book were conducted by two projects; CAS-TWAS Project, which researched flood monitoring for mitigating flood loss in frequently flooding areas based on the cooperative utilization of geospatial information technologies between China and Southeast Asian countries supported by the Chinese Academy of Sciences (CAS) and the World Academy of Sciences (TWAS) Center of Excellence on Space Technology for Disaster Mitigation (SDIM), and the Thai-China Project, which developed an application of Chinese and Thai satellite data for flood management in repeatedly flooded area supported by the 20th session of the Joint Committee on Scientific and Technical Cooperation between the governments of China and Thailand.

The research projects focused on extracting flooding areas and the diseases caused by flooding. One important thing which needs to be considered when facing flood disasters is flood loss estimation. The first part of this book discusses satellite data, which is used in the investigation and the methods for flood area identification by presenting simple, widely used method and advanced algorithm. The second part of this book is a study on waterborne diseases caused by flooding based on multi-temporal satellite imagery and a backpropagation neural network algorithm, especially infectious waterborne diseases causing diarrhea. In addition, the book also collected surveillance data for communicable diseases caused by floods based on geospatial information technologies application, which it might easily use during flooding to estimate the people who living in flooding area.

This book comprising ten chapters is a joint effort of many scientists and researchers. The results in the first part of this book were conducted by Dr. Patcharin Kamsing and Dr. Sornkitja Boonprong, and the results in the second part of this book were experimented by Dr. Peera Yomwan and Dr. Apitach Saokarn. All of the experiments performed were under the supervision of Prof. Chunxiang Cao and Dr. Min Xu. Finally, the authors would like to thank all organizations for giving support and constructive criticism. Comments and suggestions are welcome and highly appreciated.

Beijing, China   Chunxiang Cao
Beijing, China   Min Xu
Bangkok, Thailand   Patcharin Kamsing
Bangkok, Thailand   Sornkitja Boonprong
Bangkok, Thailand   Peera Yomwan
Bangkok, Thailand   Apitach Saokarn

Contents

Part I
Flooding Identification Method

Chapter 1
Geographical Characteristics of the Study Area

1.1 Characteristics of the Ayutthaya Province

The most predominant characteristics of Thailand's terrain are high mountains, an upland plateau and a central plain (Wikipedia 2014) overlooking the topography and drainage of Thailand. Northern Thailand contains many mountains, which extend along the Myanmar border down through the Kra Isthmus and the Malay Peninsula. The central part of Thailand covers a lowland area drained by the Chao Phraya River and its tributaries, the country's principal river network, which flows into the delta at the head of the Bay of Bangkok. The Chao Phraya river network drains approximately one-third of the nation's territory. In the northeastern part of the country the Khorat Plateau, a region of gently rolling low hills and shallow lakes flows into the Mekong River via the Mun River. The Mekong river network empties into the South China Sea and includes a series of canals and dams. Therefore, the Chao Phraya and Mekong river networks sustain Thailand's agricultural economy by supporting wet-rice cultivation and providing waterways for the transport of goods and people.

The case study in this book is in Ayutthaya Province consists of sixteen districts: Phra Nakhon Si Ayutthaya, ThaRuea, Nakhon Luang, Bang Sai, Bang Ban, Bang Pa-in, Bang Pahan, Phak Hai, Phachi, Lat Bua Luang, Wang Noi, Sena, Bang Sai, Uthai, MahaRat, and Ban Phraek. The area covers 2,556.6 km^2 with a population currently estimated at 787,653. Because the area has an extensive network of rivers and canals, as shown in Fig. 1.1, it is affected by flooding during the monsoon season almost every year. In addition, there are altogether 200 km of canals with a network of 1,254 canals to all the rivers in the area. Ayutthaya province, a part of Chao Phraya River basin, is one of the most severely affected provinces. The Chao Phraya River basin, located in the central part of Thailand between 13.5°–20° N and 98°–102° E(Mikhailov and Nikitina 2009),is the major economic area of Thailand that suffered the most damage in the 2011 Thailand major floods. It is the largest river basin in the region occupying about 160,000 km^2, or 30% of the country. The river's dominant flow direction is from north to south and runs through Bangkok, then empties into the Gulf of Thailand. The Chao Phraya River is divided into upper

© Higher Education Press and Springer Nature Singapore Pte Ltd. 2021
C. Cao et al., *Environmental Remote Sensing in Flooding Areas*,
https://doi.org/10.1007/978-981-15-8202-8_1

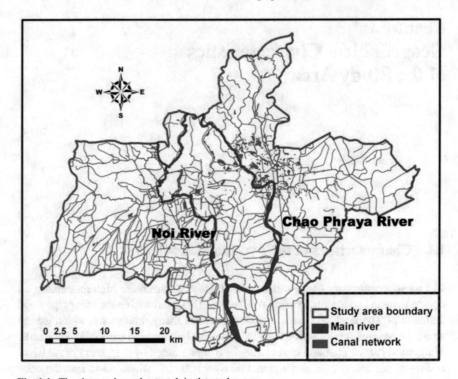

Fig. 1.1 The river and canal network in the study area

and lower river basins. Most of the terrain in the upper basin is mountainous, with 40% forest cover and 41% cultivated land. Forest areas are rapidly being converted to monoculture cash crops and built-up land, causing soil degradation and erosion in some places. The lower river basin including the river delta is a floodplain that supports intensive rice cultivation. Annual precipitation in the basin averages around 1,200 mm, and in general the major rainfall period lasts from May to late October.

1.2 Thailand Major Flood Event of 2011

Collecting the flood information from broadcast news, one can summarize the time-line of the 2011 flood affecting the Ayutthaya province and the surrounding area as shown in Fig. 1.2. From July to August 2011, the Thailand major flood began with heavy monsoon rains in northern and northeastern Thailand, causing Thailand's major dams to reach maximum capacity and cause flash floods. Afterwards, in September Typhoon Nesat and Tropical Storm Hai Yang brought further rains leading to the gradual flowing of floodwaters toward the central part of Thailand through several streams, particularly Chao Phraya River. In October, flooding hit the

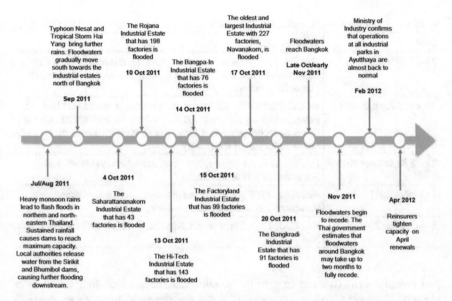

Fig. 1.2 Timeline of the 2011 Thailand major flood event affecting the main industrial estates in Ayutthaya province collected from broadcast news

central province of Ayutthaya submerging historical sites as well as five industrial estates: Saharattananakhon, Bang pa-in, Rojana, Hi-tech and Factory land.

The five estates are valued at about US$ 6.7 billion in total investment capital and employ more than 200,000 workers. In November floodwaters around Bangkok, including in Ayutthaya province, began to recede, and it took until the beginning of 2012 to fully recede. There were a number of epidemic reports during and after the 2011 Thailand major flood. The Bureau of Epidemiology (BoE), Thailand Ministry of Public Health reported the most severe infectious disease outbreaks were diarrhea, fever, pneumonia, conjunctivitis, dengue fever, leptospirosis, and hand-foot-mouth disease. The main cause of such outbreaks is people becoming infected through drinking, cooking and physical contact during the flooding periods.

According to the Meteorological Department, the 2011 annual average rainfall was 24% higher than the normal year. The genesis of this flooding disaster was five tropical cyclones (in Table 1.1) that brought continuous torrential rains and flash floods to the northern part of Thailand (GISTDA 2012).

1.3   Environmental Remote Sensing of Flooding Area

Generally flooding cases do not only impact just the economy or any infrastructure, but also impact human health. This book has two major parts. The first part is flood identification to extract the flooding area by using remote sensing data. The second

Table 1.1 The tropical cyclone affecting Thailand's weather in the year of 2011 (GISTDA 2012)

Date	Tropical cyclone
24–26 June 2011	HAIMA 1104: it moved into Nan province, resulting in torrential rains, flash floods, and landslides in many provinces in the northern part of the country.
30 July 3 August 2011	NOCK-TEN 1108: similarly to the previous, it moved into Nan Province causing the most rain in a 24-h period ever documented in 61 years; record rain fell in Nong Khai's Muang District, the northeastern part of the country.
27–28 September 2011	HAITANG 1118: The tropical storm moved into the North and Northeastern part of Thailand.
1 October 2011	NESAT 1117: Typhoon arrived at coastal areas and weakened into a depression when moving into Thailand.
5–6 October 2011	NALGAE 1119: Typhoon moved to the coast and broke up over Viet Nam.

part is waterborne diseases caused by flooding disasters, which focuses on three kinds of diseases; diarrhea, conjunctivitis, and leptospirosis. In addition, the back propagation neural network (BPNN) algorithm was implemented to predict the risk ratio of diarrheal outbreak by using some other parameters to be inputted in the algorithm. However, because of the limitation of the validation data, the implemented study area was limited to only eight districts of the Ayutthaya province. The study is a part of the indexing framework for diagnosis of environmental health by remote sensing (Fig. 1.3) developed by the Center for Applications of Spatial Information Technologies in Public Health, Aerospace Information Research Institute, Chinese Academy of Sciences (CAS). According to the framework, the main relative indices used in this study are composed of water quality index, natural disasters index, and population health index.

Recently other research explaining flood identification based on remote sensing data and geo-spatial information technologies applied for human diseases are illustrated in the next two sections to serve as basic knowledge for understanding the proposed method in this book.

1. Study on Flood Identification Based on Remote Sensing Technology

Since the late 1970s, the ability to observe information regarding bodies of water from space has proved very helpful, and created an availability of images from the Earth Resources Technology Satellite and Landsat series (Mcginnis and Rango 1975). Subsequently in past decades, very rapidly, the strength of spaceborne radar imaging for flood monitoring (e.g., penetration of clouds, functionality regardless of weather conditions, and ability to acquire data during both day and night) has become apparent (Sanyal and Lu 2004). The increasing frequency of floods caused by altered precipitation patterns triggered by climate change leads to an increasing demand for more remote sensing data (Tian et al. 2014). A frequently used and straightforward approach to deriving useful flood information from space is that of delineating a

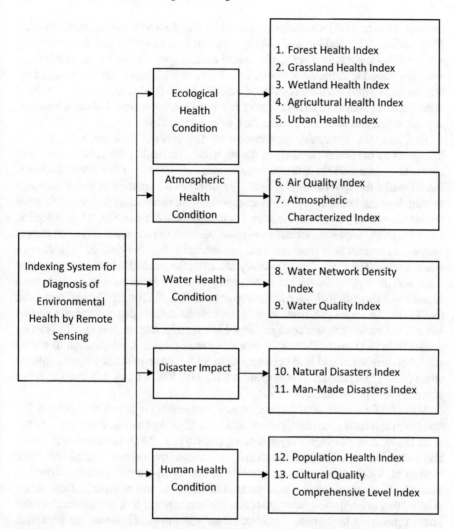

Fig. 1.3 The framework of indexing system for diagnosis of environmental health by remote sensing developed by Center for Applications of Spatial Information Technologies in Public Health under the supervision of Professor Chunxiang Cao

binary map composed of dry and flooded pixels. This procedure is applied throughout the world by numerous research teams and engineering from public and private agencies, as well as from emergency response services and governmental institutions. Imagery with the band combinations of visible and thermal bands are an appealing mode of acquisition of flood images, which have proved effective in flood mapping (Marcus and Fonstad 2008).

As water bodies can be characterized by a high reflectance in the blue wavelength and a rapid diminishment in the visible wavelengths followed by very weak signal in the near infrared (NIR) wavelengths, remote sensing data with spectral bands ranging from visible (VIS) and infrared (IR) spectra can be a powerful tool for identifying water bodies accurately from space. Based on this spectral feature, several indices were developed to delineate water bodies. These water indices aim to maximize the difference between the indices value of water bodies and other surfaces. The spectral water indices (Tran et al. 2010) encountered in scientific literature indicates how they are calculated and in which context they were originally used.

However, the systematic application of the spectral indexed techniques is hampered by persistent cloud cover during floods, particularly in small to medium-sized encampments, where floods often recede before weather conditions improve. Additionally, the applicability of optical sensors has a limitation in that it is unable to map flooding beneath vegetation canopies unlike radar imagery. Given the limitations of optical sensors to acquire flood information routinely, flood detection and monitoring seems realistically only feasible with microwave (i.e., radar) remote sensing, as microwaves penetrate cloud cover and reflect back off of smooth open water bodies. The main factors, mainly affecting the interaction between the radar wave and the water surface, are terrain roughness and instrument parameters such as radar wavelength, incidence angle and polarization. With synthetic aperture radar (SAR) data in single polarization, it is possible to detect open water quite simply when using the appropriate configuration. For example, Baghdadi et al. (2007) using ASAR/ENVISAT data (C-band) showed that the optimum configuration to distinguish between water and land corresponds to SAR data acquired at a high incidence angle (>30°) independent of polarization (HH, HV, VH, VV), or at a low incidence angle in cross-polarization (HV, VH).

Many SAR image-processing techniques exist to more or less successfully derive flood area and extent, including visual interpretation (Brivio et al. 2002; Oberstadler et al. 1997), image histogram thresholding (Brivio et al. 2002; Hostache et al. 2009; Rakwatin et al. 2013), automatic classification algorithms (Bonn and Dixon 2005; Hess et al. 1995), image texture algorithms, and multi-temporal change detection methods. Many studies have demonstrated the effectiveness of visual interpretation in extracting flooded areas from SAR data. Schumann et al. (2009b) combined several image processing techniques to produce inundation maps. However, the limitation of the repeat cycle of each SAR satellite as shown in Table 1.2 causes incoherence in monitoring dynamic events like floods. Thus, various studies attempt to combine several sources of earth observation satellites in observing flood events.

A combination of multi-sensor and multi-temporal components can mitigate the limitations of flood monitoring thanks to satellite revisiting. The combination of optical and radar images also leads to an improved damage assessment (Stramondo et al. 2006). Butenuth et al. (2011) presented a fusion technique of Gaussian mixture models and change detection methods to apply multi-sensor and multi-temporal imagery after natural disasters to data. They provided the combination of multi-sensor and multi-temporal components in a comprehensive assessment system. The combination is accomplished combining probabilities derived from the different input data. Integrating several kinds of remote sensing imagery is a key prerequisite to guarantee a fast assessment independent of the available sensor type.

Table 1.2 The missions of SAR satellites frequently used for floodplain mapping

Mission (Agency: Year of launch)	Spatial resolution (m)	Repeat cycle (days)	Band (GHz)	Polarization
ERS-2 (ESA: 1995)	25	35	5.3	VV
Radarsat-I (CSA: 1995)	8–100	24	5.3	HH
ENVISAT (ESA: 2002)	12.5–1000	35	5.3	VV-VH, VV-HV
ALOS (JAXA: 2006)	7–100	46	1.3	Full
COSMO-SkyMed (ASI: 2007)	15–100	16	9.6	Dual
TerraSAR-X (DLR: 2007)	1–16	11	9.6	Full
Radarsat-2 (MDA: 2007)	3–100	24	5.3	Full

Meanwhile, a proposed modular system is able to deal with varying data sources embedding all obtainable information to ensure the transferability of the improved strategy and methods. Moreover, the integration of different imagery from different points in time provides several advantages compared to current solutions. Multitemporal images provide the opportunity to monitor a natural disaster chronologically during a period of time, not only at a specific time point. Rakwatin et al. (2013) combined SAR imagery, optical satellite imagery and a digital elevation model (DEM) with water level data from gauge stations to map the area flooded and to estimate water volume in near real time to support decision-making for flood relief operations. They provided difficulties of encountered when dealing with different kinds of spatial data and different application techniques and revealed problems, which included inconsistent acquisition schedules for different satellites, different image resolutions and different data acquisition modes, i.e. ScanSAR Wide and Wide modes. Their work is underway to improve the satellite image acquisition planning and DEM accuracy and increase the number of gauge stations in the flood-affected areas in order to improve the accuracy, reliability and usefulness of geo-informatics data for future disaster management.

2. Study on Geo-Spatial Information Technologies applied for Human Diseases

Remote sensing (RS) has been used to detect and analyze environmental factors for several decades (Cao et al. 2010; Herbreteau et al. 2007; Schumann et al. 2009a). However, RS has not been used very often for studying the dynamics of environmentally dependent diseases, such as waterborne diseases, and has begun this research trend only as recently as 2007: leptospirosis, flooding, and water poisoning (Herbreteau et al. 2007). In recent years, the improvement of satellite sensors has led a number of researchers to utilize their data for assessing the risk of waterborne disease

(Cao et al. 2012; Lleo 2009). Various studies have applied earth-observing satellites and geographic information system (GIS) modeling for the surveillance and modeling of waterborne disease. For example, Constantin de Magny et al. (2008) developed a prediction model for cholera by utilizing satellite sensors to measure chlorophyll concentration and sea-surface temperature. Ford et al. (2009)used satellite images of environmental changes to model cholera outbreaks. Tran et al. (2010)analyzed satellite images for water detection and focused on the main variables that influence the survival of avian influenza viruses in water. However, the first relevant studies of disease risk in flood disasters only appeared in 2012 (Kazama et al. (2012), Yomwan et al. (2012), and examined the use of spatial-information technologies for assessing the risk of waterborne infectious disease. Because their studies integrated flood parameters into the quantitative microbial risk assessment (QMRA) (Howard et al. 2006a, b), they needed a directly measured pathogen parameter, which requires a complicated and time-consuming laboratory analysis and cannot easily be applied to a large number of spatially distributed samples.

1.4 Summary

The aspect of the study area mentioned in the first part includes the geography and the impact of monsoon season in the study area. The study area is located in central Thailand and is important for both an agriculture and industry of Thailand. Thailand flooding in 2011 caused damages to many provinces, especially Ayutthaya Province. This chapter collects the time-series data of typical cyclones and affected of flooding during that time. Remote sensing technology has become a vital tool for solving the problems caused by floods; for example, the waterborne diseases resulting from flooding. Many researchers have conducted research based on different datasets, however, this topic remains as a popular issue because of the different characteristic in different study area.

References

Baghdadi N, Pedreros R, Lenotre N, Dewez T, Paganini M (2007) Impact of polarization and incidence of the ASAR sensor on coastline mapping: example of Gabon. Int J Remote Sens 28(17):3841–3849

Bonn F, Dixon R (2005) Monitoring flood extent and forecasting excess runoff risk with RADARSAT-1 data. Nat Hazards 35(3):377–393

Brivio PA, Colombo R, Maggi M, Tomasoni R (2002) Integration of remote sensing data and GIS for accurate mapping of flooded areas. Int J Remote Sens 23(3):429–441

Butenuth M, Frey D, Nielsen AA, Skriver H (2011) Infrastructure assessment for disaster management using multi-sensor and multi-temporal remote sensing imagery. Int J Remote Sens 32(23):8575–8594

Cao CX, Xu M, Chang CY, Xue Y, Zhong SB, Fang LQ, Cao WC, Zhang H, Gao MX, He QS et al (2010) Risk analysis for the highly pathogenic avian influenza in Mainland China using meta-modeling. Chin Sci Bull 55(36):4168–4178

Cao CX, Xu M, Chen W, Tian R (2012) A framework for diagnosis of environmental health based on remote sensing. Land Surf Remote Sens 8524

Constantin de Magny G, Murtugudde R, Sapiano MRP, Nizam A, Brown CW, Busalacchi AJ, Yunus M, Nair GB, Gil AI, Lanata CF et al (2008) Environmental signatures associated with cholera epidemics. Proc Nat Acad Sci USA 105(46):17676–17681

Ford TE, Colwell RR, Rose JB, Morse SS, Rogers DJ, Yates TL (2009) Using satellite images of environmental changes to predict infectious disease outbreaks. Emerg Infect Dis 15(9):1341–1346

GISTDA (2012) The world of water

Herbreteau V, Salem G, Souris M, Hugot J-P, Gonzalez J-P (2007) Thirty years of use and improvement of remote sensing, applied to epidemiology: from early promises to lasting frustration. Health Place 13(2):400–403

Hess LL, Melack JM, Filoso S, Wang Y (1995) Delineation of inundated area and vegetation along the amazon floodplain with the sir-c synthetic-aperture radar. IEEE Trans Geosci Remote Sens 33(4):896–904

Hostache R, Matgen P, Schumann G, Puech C, Hoffmann L, Pfister L (2009) Water level estimation and reduction of hydraulic model calibration uncertainties using satellite SAR images of floods. IEEE Trans Geosci Remote Sens 47(2):431–441

Howard G, Pedley S, Tibatemwa S (2006a) Quantitative microbial risk assessment to estimate health risks attributable to water supply: can the technique be applied in developing countries with limited data? J Water Health 4(1):49–65

Howard G, Pedley S, Tibatemwa S. (2006b) Quantitative microbial risk assessment to estimate health risks attributable to water supply: can the technique be applied in developing countries with limited data? Water Health

Kazama S, Aizawa T, Watanabe T, Ranjan P, Gunawardhana L, Amano A (2012) A quantitative risk assessment of waterborne infectious disease in the inundation area of a tropical monsoon region. Sustain Sci 7(1):45–54

Lleo MD (2009) Application of space technologies to the surveillance and modelling of waterborne diseases. Trop Med Int Health 14:23–23

Marcus WA, Fonstad MA (2008) Optical remote mapping of rivers at sub-meter resolutions and watershed extents. Earth Surf Proc Land 33(1):4–24

Mcginnis DF, Rango A (1975) Earth resources satellite systems for flood monitoring. Geophys Res Lett 2(4):132–135

Mikhailov VN, Nikitina OI (2009) Hydrological and morphological processes in the Chao Phraya Mouth Area (Thailand) and their anthropogenic changes. Water Resour 36(6):613–624

Oberstadler R, Honsch H, Huth D (1997) Assessment of the mapping capabilities of ERS-1 SAR data for flood mapping: a case study in Germany. Hydrol Process 11(10):1415–1425

Rakwatin P, Sansena T, Marjang N, Rungsipanich A (2013) Using multi-temporal remote-sensing data to estimate 2011 flood area and volume over Chao Phraya River basin, Thailand. Remote Sens Lett 4(3):243–250

Sanyal J, Lu XX (2004) Application of remote sensing in flood management with special reference to monsoon Asia: a review. Nat Hazards 33(2):283–301

Schumann G, Di Baldassarre G, Bates PD (2009a) The Utility of Spaceborne Radar to Render Flood Inundation Maps Based on Multialgorithm Ensembles. IEEE Trans Geosci Remote Sens 47(8):2801–2807

Schumann G, Bates PD, Horritt MS, Matgen P, Pappenberger F (2009b) Progress in integration of remote sensing-derived flood extent and stage data and hydraulic models. Rev Geophys 47

Stramondo S, Bignami C, Chini M, Pierdicca N, Tertulliani A (2006) Satellite radar and optical remote sensing for earthquake damage detection: results from different case studies. Int J Remote Sens 27(20):4433–4447

Tian R, Cao C, Peng L, Ma G, Bao D, Guo J, Yomwan P (2014) The use of HJ-1A/B satellite data to detect changes in the size of wetlands in response into a sudden turn from drought to flood in the middle and lower reaches of the Yangtze River system in China. Geomatics, Natural Hazards and Risk:1–21

Tran A, Goutard F, Chamaille L, Baghdadi N, Lo Seen D (2010) Remote sensing and avian influenza: a review of image processing methods for extracting key variables affecting avian influenza virus survival in water from Earth Observation satellites. Int J Appl Earth Obs Geoinf 12(1):1–8

Wikipedia (2014) Geography of Thailand[Internet]. http://en.wikipedia.org/wiki/Geography_of_Tha150iland

Yomwan P, Cao CX, Rakwatin P, Apaphant P (2012) The risk analysis for infectious disease outbreaks in flood disaster based on spatial information technologies. In: 2012 IEEE International Geoscience and Remote Sensing Symposium (IGARSS):7244–7247

Chapter 2
Datasets and Data Preparation

2.1 Remote Sensing Datasets and Preprocessing

1. Remote Sensing Datasets

Remotely sensed imagery plays an important role in providing up-to-date information that is spatially accurate and captures a wide area with frequent, repeated observations. To efficiently monitoring flood events, we need remote sensing data as much as possible. However, because of the limitation of the revisiting of each earth observation (EO) satellite, using multi-sensor and multi-temporal EO satellite imageries will lead us to more frequently update flood situations than using only one EO satellite data. The EO satellite data in this study are composed of THEOS, HJ-1A/B and Radarsat-2 imageries.

THEOS, also known as Thaichote, is an earth observation mission from Thailand developed at EADS Astrium SAS in Toulouse, France. In July 2004, EADS Astrium SAS signed a contract for delivery of THEOS with GISTDA (Geo-Informatics and Space Technology Development Agency) of Bangkok, Thailand. The characteristics of THEOS satellite is illustrated in Table 2.1.

According to Table 2.1, THEOS has two sensors for panchromatic and multispectral mode. The panchromatic strip is 22-km width, while the multispectral strip is approximately 90-km. width at nadir. An image is defined as a square portion of the image strip. An image strip can be acquired consistently with a length of up to 10 min that is an equivalent of about 4,000 km. THEOS oblique viewing capability serves for the imaging of any area within a 1,000 km swath (for 30° roll). During a given cycle, the viewing frequency for a given point can be enhanced by using oblique viewing. The frequency relies on latitude. For example, over Thailand a given area can be imaged 9 times during the same 26-day orbital cycle. This means 126 yearly revisits and an average of 3 days, with an interval ranging from a minimum of 1 day up to a maximum of 5 days.

© Higher Education Press and Springer Nature Singapore Pte Ltd. 2021
C. Cao et al., *Environmental Remote Sensing in Flooding Areas*,
https://doi.org/10.1007/978-981-15-8202-8_2

Table 2.1 THEOS satellite characteristics

THEOS ssensors	Panchromatic	Multispectral
Wavelength (μm)	P: 0.45–0.90	B0 (blue): 0.45–0.52 B1 (green): 0.53–0.60 B2 (red): 0.62–0.69 B3 (near infrared): 0.77–0.90
Resolution (m)	2	15
Swath width (km)	22 (nadir)	90 (nadir)
Number of pixels	12,000	6,000
Access corridor width (km)	1,000	1,100

Huanjing means environment in Chinese, and the small satellite constellation Huanjing 1 (HJ-1) is one of satellite missions for environment and disaster monitoring and forecasting (Jiang et al. 2013). The HJ-1 satellite constellation consists of a number of application systems, small satellites and ground systems. In the first stage, the HJ-1 constellation consists of two optical satellites, HJ-1A and HJ-1B, and one radar satellite, HJ-1C. On September 6, 2008, the HJ-1A/B satellites were successfully launched at Taiyuan Satellite Launch Center in Shanxi Province. They are sun-synchronous circular orbit satellites with an orbital altitude of 649 km. The single charge-coupled device (CCD) imagery width is 360-km, and the two satellite constellations provide an observation re-visit cycle of 48 h. The CCD cameras are composed of four bands, including blue, green, red, and near-infrared spectrums, with a spatial resolution of 30-band and spectral range of 0.43–0.9 μm. The HJ-1A/B data are freely available to the public.

HJ-1A/B are a new generation of small Chinese civilian earth observation optical remote sensing satellites. They are distributed with a phase difference of 180° in the same orbital plane to increase the temporal resolution of earth observations and to obtain mosaic images from the two satellites. The payload of HJ-1A includes a data transmission system, two wide-coverage multispectral CCD cameras for field splicing to realize an image width of 720 km, one hyper-spectral imager (HSI), and a Ka communications tests system. The payload of HJ-1B includes two CCD cameras identical to those in HJ-1A, one infrared camera, and a data transmission subsystem. The main sensors and the parameters of HJ-1A/B are shown in Table 2.2. With high revisit frequency, wide imaging coverage, and good spatial resolution, the HJ-1A/B imagery has been widely used, mainly in China, for retrieving land surface parameters, monitoring water quality, air quality, and various natural disasters

With the advantages of SAR data for water-body extraction and cloud penetration, the SAR data used in this study is a time series of the Radarsat-2 data set. Radarsat-2, the second in a series of Canadian space borne SAR satellites launched in 2007, has a single-sensor polarimetric C-band SAR (5.405 GHz) with multiple polarization modes (HH, HV, VV, and VH) and has a sun-synchronous orbit at an altitude of

Table 2.2 HJ-1A/B satellite characteristics (Jiang et al. 2013)

Satellite	Payload	Band no.	Spectral range (μm)	Nadir spatial resolution (m)	Swath width (km)	Side-looking ability (degree)	Repetition cycle (h)	Date transmission rate (Mbps)
HJ-1A/B	Multispectral CCD camera	1	0.43–0.52	30	360 (for two ≥720)	–	48	120
		2	0.52–0.60					
		3	0.63–0.69					
		4	0.76–0.90					
HJ-1A	Hyperspectral imager	–	0.45–0.95 (110–128 bands)	100	≥50	±30		
HJ-1B	Infrared multispectral camera	5	0.75–1.10	150	720	–	96	60
		6	1.55–1.75					
		7	3.50–3.90					
		8	10.5–12.5	300				

798 km with a 6 PM ascending node and a 6 AM descending node. Radarsat-2 has the capability of routine left- and right-looking operations (the right-looking mode for the default operation and the left-looking mode for improved monitoring efficiencies in case of emergency imaging requests and for regions not covered in the right-looking mode, e.g. Antarctica). The resolutions of Radarsat-2 imagery on several modes are expressed in Table 2.3.

The multi-satellite imageries used in this study include:

(1) Six Radarsat-2 scenes of 50-m resolution with the ScanSAR narrow mode acquired on (i) 9 September 2011, (ii) 3 October 2011, (iii) 21 October 2011, (iv) 14 November 2011, (v) 4 December 2011, and (vi) 28 December 2011;
(2) Three THEOS scenes of 15-m resolution with the multispectral sensor mode acquired on (i) 4 November 2011, (ii) 1 December 2011, and (iii) 13 December 2009;
(3) Three HJ-1A/B scenes of 30-m resolution with the multispectral sensor mode acquired on (i) 4 August 2011, (ii) 9 November 2011, and (iii) 24 January 2012.

2. Image Preprocessing

Operations of image preprocessing, sometimes referred to as image restoration and geo-rectification, are proposed to correct for sensor- and platform-specific radiometric and geometric distortions of the image. The process attempts to produce a

Table 2.3 Radarsat-2 satellite characteristics and modes. Beam mode name, swath width, swath coverage, and nominal resolution

	Beam mode	Nominal swath width (km)	Swath coverage to left of right of ground track (km)	Approximate resolution: Rng (m) × Az (m)
Radarsat-1 modes with selective polarization Transmit H or V Receive H or V or (H and V)	Standard	100	250–750	25 × 28
	Wide	150	250–650	25 × 28
	Low incidence	170	125–300	40 × 28
	High incidence	70	750–1000	20 × 28
	Fine	50	525–750	10 × 9
	ScanSAR wide	500	250–750	100 × 100
	ScanSAR narrow	300	300–720	50 × 50
Polarimetry Transmit H and V on alternate pulses Receive H and V on every pulse	Standard QP	25	250–600	25 × 28
	Fine QP	25	400–600	11 × 9
Selective single polarization Transmit H or V Receive H or V	Multiple fine	50	400–750	11 × 9
	Ultra-fine wide	20	400–550	3 × 3

corrected image that is as close as possible, both geometrically and radiometrically, to the radiant energy characteristics of the original scene. Radiometric corrections may be necessary due to variations in scene illumination, viewing geometry, atmospheric conditions, sensor noise and response. Each of these will vary depending on the specific sensor and platform used to acquire the data and the conditions during data acquisition. In addition, it may be desirable to convert and/or calibrate the data to known (absolute) radiation or reflectance units to facilitate comparison between data.

In general, before image analysis initial processing on the raw scene is carried out to correct for any distortion due to the characteristics of the imaging system and imaging conditions. The standard correction procedures can be carried out by the ground station operators before the data is delivered to the end-user depending on the user's requirement. These procedures include radiometric correction to correct for uneven sensor response over the whole image and geometric correction to correct for geometric distortion due to Earth's rotation and other imaging conditions (such as oblique viewing). The image may also be transformed to conform to a specific map projection system. Moreover, if accurate geographical location of an area on the image needs to be determined, the image is registered to a precise map (geo-referencing) by using ground control points (GCPs).

The pre-processing of remotely sensed images is generally divided into two distinct processes: image enhancement or radiometric correction, and georeferencing or geometric correction.

2.1.1 Image Enhancement or Radiometric Correction

The image enhancement process aims to correct the raw image and make it more suitable to the capabilities of human vision. Regardless of the extent of digital intervention, visual analysis invariably plays a very strong role in all aspects of remote sensing. Image enhancement techniques, such as grey level stretching to increase the contrast and spatial filtering for enhancing the edges, can improve visual appearance of the objects in the image and proves very useful for visual interpretation. While the range of image enhancement techniques is broad, the following fundamentals form the backbone of this area including in this study.

1. Contrast Stretch

Digital sensors of EO satellites have a wide range of output values to accommodate the strongly varying reflectance values that can be found in different environments. However, for a single environment, it is often to the case that only a narrow range of values will occur over most areas. Grey level distributions thus tend to be fairly skewed. Contrast manipulation procedures are thus essential to most visual analyses. Note that the values of the image are quite skewed. From Fig. 2.1, the right image shows the same image band after a linear stretch between values 12 and 60 has been

Fig. 2.1 A contrast stretch approach for image enhancement

applied. This type of contrast enhancement may be performed interactively while the image is displayed. This is normally used for visual analysis only—original data values are used in numeric analyses. New images with stretched values are produced with modules about stretching.

2. Composite Generation

For visual analysis, color composites make fullest use of the capabilities of the human eye. Depending upon the graphics system in use, composite generation ranges from simply selecting the bands to use to more involved procedures of band combination and associated contrast stretch. In this study, we utilized the composite generation in combination of multispectral bands of THEOS and HJ-1A/B imageries.

3. Digital Filtering

One of the most intriguing capabilities of digital analysis is the ability to apply digital filters. Filters can be used to provide edge enhancement (sometimes called crispening) to remove image blur and to isolate lineaments and directional trends to mention just a few.

To reduce the speckle noise of SAR images, the most widely used adaptive filters based on the spatial domain include the Lee, Frost, Enfrost, Kuan, Median, and Gamma filters (Matgen et al. 2007). The 5 × 5 Kuan filter was applied in the study by using trial and error with visualization in order to reduce the speckle noise for the Radarsat-2 images (Gupta and Gupta 2007). Based on the criterion of minimum mean square error, the Kuan filter applies a spatial filter to each pixel that is replaced with a value calculated based on the local statistics and can reduce speckle while preserving edges by transforming the multiplicative noise model into an additive noise model (Kuan et al. 1985).

2.1.2 Georeferencing or Geometric Correction

Ordinarily, most remotely sensed imagery is served with some level of geo-referencing information, which locates the image in a ground coordinate system.

There are generally three levels of geo-referencing, each corresponding to a different geometric accuracy (Karen Schuckman 2014).

(1) Level 1: uses positioning information obtained directly from the sensor and platform to roughly geo-locate the remotely sensed scene on the ground. This level of geo-referencing is efficient to gain geographic context and support visual interpretation of the data. However, it is not often not accurate sufficient to support robust image or GIS analysis that requires a combination of the remotely sensed dataset with other layers.

(2) Level 2: uses a digital elevation model (DEM) to eliminate relief displacement due to variation in the height of the terrain. This enhances the relative spatial accuracy of the data that means distances measured between points within the geo-corrected image will be more accurate than the level 1, particularly in scenes containing robust elevation changes. Generally, the DEM is obtained from another source, and the spatial accuracy of the Level 2 image will rely on the accuracy of the DEM.

(3) Level 3: uses a DEM and ground control points (GCPs) to most accurately geo-referenced the image on the ground. In addition to the DEM, ground control points must be obtained from another source, and the accuracy of the Level 3 image will rely on the accuracy of the ground control points. Level 3 processing is frequently required in order to provide the most accurate overlays of remotely sensed data sets and other relevant GIS data.

For this study, a series of flooded area map was extracted from SAR and multi-spectral imageries. SAR has significant advantages for the detection of water bodies and can penetrate clouds. Radarsat is in operational use for flood monitoring in many countries (Brisco et al. 2008). It has been shown to accurately assess and clarify inundated areas. Moreover, its ability to penetrate clouds is very important for monitoring floods during the rainy season in monsoon countries (Hoque et al. 2011). Radarsat-2 is the second in a series of Canadian spaceborne SAR satellites that provides several improvements over Radarsat-1, such as additional beam modes, higher resolution, multi-polarization, and more-frequent revisits.

In the beginning of this study, we obtain the level 1 remote sensing data sets including Radarsat-2, HJ-1A/B and THEOS imageries, acquired from August 2011 to January 2012. The multi-temporal remote sensing images were manually orthorectified with topographic base maps, and the nearest-neighbor method was used to preserve original values in the re-sampling process. Each image was orthorectified using at least eight ground control points that were manually selected in order to correct the image geometry to that of the to pographic base maps. The base maps were derived from rectified aerial photomap with 50 cm spatial resolution. The Universal Transverse Mercator zone 47 was defined as the image-to-map projection. The acceptable threshold of the Root Mean Square (RMS) error was set to one pixel due to limited human resources and time constraints. Figure 2.2 provides an example of geo-referencing for a Radarsat-2 image and its comparison of the image before and after preprocessing.

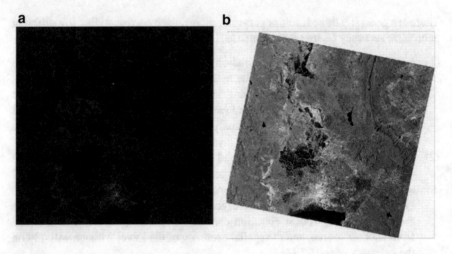

Fig. 2.2 Comparison of a Radarsat-2 image before (a) and after (b) pre-processing which includes radiometric and geometric correction

2.2 Flood Water Quality Data

In a flood disaster, fecal coliform bacteria, or *Escherichia coli*, are common causes of diarrhea because the flooding washes fecal material from human habitats, causing increased transmission of bacterial infection. To investigate waterborne diseases, almost all studies have attempted to determine the presence of pathogens such as bacteria and viruses in contaminated water based on quantitative microbial risk assessment (QMRA) (Howard et al. 2006). However, such studies rarely report to the spatial distribution of the risk, because QMRA requires a complicated and time-consuming laboratory analysis and so cannot easily be applied to a large number of spatially distributed samples. Time-consuming laboratory testing is required to determine the amount of pathogens in each water sample.

Dissolved oxygen (DO) is a simpler indicator of water quality that has been very widely used to assess water quality (Kannel et al. 2007). DO values can be used to indicate the degree of pollution by organic matter and the level of self-purification of the water. A number of studies expressed a close relationship between DO and diarrheal pathogens, such as fecal coliform bacteria (*E. coli*) and *Vibrio cholera* (Islam et al. 2007; Kersters et al. 1995; Massoud 2012; Osode and Okoh 2010). In particular, Osode and Okoh (2010) revealed that DO negatively correlated with *E. coli* densities ($P < 0.001$). The use of DO instead of parameters of the pathogens as an indicator of the water-quality surveillance system, as in many countries including Thailand (Boonsoong et al. 2010), can more comprehensively model the risk of waterborne disease for spatial analysis with up-to-date water quality data.

DO is an important indicator of river health (the ecological condition of a river) and is used by regulators as part of the classification scheme for good chemical status (Williams and Boorman 2012). Therefore, researchers have frequently used

DO to evaluate water quality (Kannel et al. 2007). In addition, DO have a close relationship to fecal coliform bacteria. These bacteria are oxidase negative, therefore if DO decreases, then coliform bacteria frequently increase and vice versa.

In the study, DO was used as an input factor governing the risk of diarrheal infection of people in inundated areas. The (DO) values were derived from 186 floodwater samples collected by the Pollution Control Department of Thailand during the flood. Each water sample includes water quality indicators and relative details that can be divided into 3 parts.

(1) Water chemical indicators: DO (mg/L), pH, and Salinity (ppt)
(2) Water physical indicators: Water Temp. ($^\circ$C), Conduct (mS), Turbidity (NTU), Water Color, and Smell
(3) Floodwater variables: Flood Depth, Flow, and Water Type
(4) Other details: ID., Measurement Time, Location, Land Use, Observer, and Remarks.

2.3 Morbidity Data

The weekly surveillance reports of patients in Ayutthaya obtained from the Bureau of Epidemiology (BoE), Thailand Ministry of Public Health were employed to measure the risk of outbreak and serve as the reference for the model predictions in this study. The morbidity dataset of waterborne diseases used in the study for investigating its relationship with flood disasters are composed of diarrhea, conjunctivitis, and leptospirosis. The reported cases of notifiable diarrhea, conjunctivitis, and leptospirosis by week for eight districts in the study area in 2011 are expressed in Table 2.4, Table 2.5, and Table 2.6 respectively, while Table 2.7 provides the definition of weekly surveillance reports of their tables determined by the Bureau of Epidemiology (BoE), Thailand Ministry of Public Health.

2.4 Summary

The remote sensing datasets in this study include a series of Radarsat-2, HJ-1A/B and THEOS imageries. The pre-processing of remotely sensed images is generally composed of image enhancement and geo-referencing. Dissolved oxygen (DO) was used as an input factor governing the infection risk of waterborne diseases of people in inundated areas, because it is a common indicator of water quality that has been widely used in many countries and has a close relationship between DO and waterborne pathogens, such as fecal coliform bacteria, *E. coli*, and *Vibrio cholerae*.

The weekly surveillance patient reports of three waterborne diseases (diarrhea, conjunctivitis and leptospirosis) were employed to measure the risk of outbreak and serve as the reference for the model predictions in this study.

Table 2.4 Reported cases of notifiable diarrhea by week for eight districts in the study area in 2011

District	Week (cases)																										
	01	02	03	04	05	06	07	08	09	10	11	12	13	14	15	16	17	18	19	20	21	22	23	24	25	26	27
Phra Nakhon Si Ayudhya	139	142	131	125	140	129	88	104	139	110	120	87	83	85	87	84	94	80	83	88	100	102	120	131	103	106	120
Bang Chai	28	27	27	37	27	29	26	25	16	22	17	10	13	10	13	15	17	24	22	19	30	20	19	22	22	23	21
Bang Ban	6	9	19	21	13	16	14	12	17	21	8	5	10		7	4	8	4	9	3	5	5	9	6	7	11	6
Bang Pa-in	65	71	83	70	53	57	61	49	49	39	32	25	42	49	28	47	36	39	29	58	48	44	45	35	39	34	26
Bang Pahan	47	47	49	62	32	32	22	20	28	29	23	20	17	21	16	15	13	16	19	10	21	17	25	11	18	23	15
Sena	52	47	40	39	46	45	53	38	40	46	41	36	44	23	24	33	6	12	9	5	8	7	7	10	13	5	5
Bang Sai	4	16	10	8	19	9	4	22	9	11	7	8	8	7	5	11	11	10	9	3	7	3	8	10	12	10	4
Uthai	25	40	25	33	42	45	35	24	24	32	40	24	25	15	5	15	11	16	9	9	16	19	14	21	10	27	49
SUM	366	399	384	395	372	362	303	294	322	310	288	215	242	210	185	224	196	201	189	195	235	217	247	246	224	239	246

(continued)

Table 2.4 (continued)

District	Week (cases)																									SUM
	28	29	30	31	32	33	34	35	36	37	38	39	40	41	42	43	44	45	46	47	48	49	50	51	52	
Phra Nakhon Si Ayudhya	90	78	69	59	82	77	110	45	99	107	124	129	88	62	89	64	118	126	150	177	168	141	110	105	88	5475
Bang Chai	23	15	24	14	17	13	19	9	22	15	16	33	15	14	13	7	15	22	30	21	34	28	21	23	14	1058
Bang Ban	5	9	3	5		7	11	11	8	16	11	18	23	12	14	7	9	13	23	8	13	8	15	9	4	517
Bang Pa-in	29	29	23	34	34	32	25	10	39	44	50	67	16	38	78	61	45	54	50	48	58	61	36	49	40	2303
Bang Pahan	17	8	16	16	14	6	11	14	16	12	16	18	7		7	4	8	12	28	26	35	17	11	6	9	1002
Sena	10	12	31	7	23	20	20	34	11	23	46	43	27	43	57	44	63	70	76	71	35	45	38	58	10	1651
Bang Sai	9	7	5	10	11	8	1	6	11	3	4	1	8	7	8	11	16	27	11	14	15	4	4	8	5	459
Uthai	34	31	26	17	7	12	23	16	36	28	21	33	36	7	20	31	34	39	55	53	32	37	28	36	25	1367
SUM	217	189	197	162	188	175	220	145	242	248	288	342	220	183	286	229	308	363	423	418	390	341	263	294	195	13832

Table 2.5 Reported cases of notifiable conjunctivitis by week for eight districts in the study area in 2011

District	Week (cases)																										
	01	02	03	04	05	06	07	08	09	10	11	12	13	14	15	16	17	18	19	20	21	22	23	24	25	26	27
Phra Nakhon Si Ayudhya	12	12	12	16	16	22	12	14	15	13	12	12	14	13	6	10	11	7	9	6	8	4	15	12	7	10	6
Bang Chai	6	5	6	4	1	7	10	4	4	12	5	3	2	2	1	6	3	3	4	5	4	4		3	1	2	2
Bang Ban		1	1	2	3		2	1		5			1			2	4				1		1			2	4
Bang Pa-in	12	10	11	12	8	15	13	7	8	11	5	7	6	10	8	9	12	11	4	5	7	4	7	7	8	10	5
Bang Pahan	1	3	2		2	1		1	1	1		1	2			1			1				1				
Sena	8	13	15	8	11	11	4	7	8	9	9	13	5	4	5	5	1					1	2	1	1		4
Bang Sai	3	4	3	2	1	4	7		1		1	3	3	1	1	1				2	1		2				2
Uthai		1		1	2	1		1	6	6	8	4	4			3	1		1			1				1	4
SUM	42	49	50	45	44	61	48	35	43	57	40	43	37	30	21	37	32	21	19	18	21	14	28	23	17	25	27

District	Week (cases)																									SUM
	28	29	30	31	32	33	34	35	36	37	38	39	40	41	42	43	44	45	46	47	48	49	50	51	52	
Phra Nakhon Si Ayudhya	8	13	10	8	9	12	18	12	11	10	11	20	11	8	12	9	23	20	22	8	12	15	11	12	3	614
Bang Chai	3	5	2	1	2	1	1	2	4	5	1	1	3	3	1	1	1	1	4	4		5	2	1		158
Bang Ban	1	1	1	2	1	2	2			1	5	4	1	3	1	1	3	5	1	3	2		2			68
Bang Pa-in	6	12	7	6	7	2	3		6	12	5	7	2	7	14	6	8	8	10	2	4	9	9	5	6	395
Bang Pahan												1	1	1								1		1		22
Sena			3	3	7	5	6	8	1	9	6	7	5	2	2	2	7	4	7	10	2	2	2	6	1	252
Bang Sai				1	1	1			2				1	1	1	1	1	4	3	3	3	1		1	1	62
Uthai	5	3	2		2				1	2	2	5	4		4	3	4	5	3	2	2	3	3	2	4	103
SUM	23	34	25	20	27	21	30	22	25	39	30	45	28	21	35	23	47	47	50	29	25	32	27	27	15	1647

Table 2.6 Reported cases of notifiable leptospirosis by week in the study area in 2011 (only three districts had the reported leptospirosis cases)

District	Week (cases)																										
	01	02	03	04	05	06	07	08	09	10	11	12	13	14	15	16	17	18	19	20	21	22	23	24	25	26	27
Phra Nakhon Si Ayudhya																			1								
Bang Pa-in																								1			
Uthai																					1	1					
SUM																			1		1	1		1			

District	Week (cases)																									SUM
	28	29	30	31	32	33	34	35	36	37	38	39	40	41	42	43	44	45	46	47	48	49	50	51	52	
Phra Nakhon Si Ayudhya																										1
Bang Pa-in															2				2							5
Uthai												1						1								4
SUM												1			2			1	2							10

Table 2.7 The definition of weekly surveillance reports from the Bureau of Epidemiology of Thailand in 2011 using in this study

Week	Date	Week	Date
1	2-Jan–8-Jan	27	3-Jul–9-Jul
2	9-Jan–15-Jan	28	10-Jul–16-Jul
3	16-Jan–22-Jan	29	17-Jul–23-Jul
4	23-Jan–29-Jan	30	24-Jul–30-Jul
5	30-Jan–5-Feb	31	31-Jul–6-Aug
6	6-Feb–12-Feb	32	7-Aug–13-Aug
7	13-Feb–19-Feb	33	14-Aug–20-Aug
8	20-Feb–26-Feb	34	21-Aug–27-Aug
9	27-Feb–5-Mar	35	28-Aug–3-Sep
10	6-Mar–12-Mar	36	4-Sep–10-Sep
11	13-Mar–19-Mar	37	11-Sep–17-Sep
12	20-Mar–26-Mar	38	18-Sep–24-Sep
13	27-Mar–2-Apr	39	25-Sep–1-Oct
14	3-Apr–9-Apr	40	2-Oct–8-Oct
15	10-Apr–16-Apr	41	9-Oct–15-Oct
16	17-Apr–23-Apr	42	16-Oct–22-Oct
17	24-Apr–30-Apr	43	23-Oct–29-Oct
18	1-May–7-May	44	30-Oct–5-Nov
19	8-May–14-May	45	6-Nov–12-Nov
20	15-May–21-May	46	13-Nov–19-Nov
21	22-May–28-May	47	20-Nov–26-Nov
22	29-May–4-Jun	48	27-Nov–3-Dec
23	5-Jun–11-Jun	49	4-Dec–10-Dec
24	12-Jun–18-Jun	50	11-Dec–17-Dec
25	19-Jun–25-Jun	51	18-Dec–24-Dec
26	26-Jun–2-Jul	52	25-Dec–31-Dec

References

Boonsoong B, Sangpradub N, Barbour MT, Simachaya W (2010) An implementation plan for using biological indicators to improve assessment of water quality in Thailand. Environ Monit Assess 165(1–4):205–215

Brisco B, Touzi R, van der Sanden JJ, Charbonneau F, Pultz TJ, D'Iorio M (2008) Water resource applications with RADARSAT-2—a preview. Int J Digital Earth 1(1):130–147

Eastman JR (2001) Introduction to remote sensing and image processing: guid to GIS and image processing. 17–34

Gupta KK, Gupta R (2007) Despeckle and geographical feature extraction in SAR images by wavelet transform. Isprs J Photogr Rem Sens 62(6):473–484

Hoque R, Nakayama D, Matsuyama H, Matsumoto J (2011) Flood monitoring, mapping and assessing capabilities using RADARSAT remote sensing, GIS and ground data for Bangladesh. Nat Hazards 57(2):525–548

Howard G, Pedley S, Tibatemwa S (2006) Quantitative microbial risk assessment to estimate health risks attributable to water supply: can the technique be applied in developing countries with limited data? J Water Health 4(1):49–65

Islam MS, Brooks A, Kabir MS, Jahid IK, Islam MS, Goswami D, Nair GB, Larson C, Yukiko W, Luby S (2007) Faecal contamination of drinking water sources of Dhaka city during the 2004 flood in Bangladesh and use of disinfectants for water treatment. J Appl Microbiol 103(1):80–87

Jiang B, Liang S, Townshend JR, Dodson ZM (2013) Assessment of the radiometric performance of Chinese HJ-1 Satellite CCD instruments. IEEE J Sel Top Appl Earth Obs Remote Sens 6(2):840–850

Kannel PR, Lee S, Lee YS, Kanel SR, Khan SP (2007) Application of water quality indices and dissolved oxygen as indicators for river water classification and urban impact assessment. Environ Monit Assess 132(1–3):93–110

Karen Schuckman JAD (2014) Exploring imagery and elevation data in GIS applications. e-Education Institute, College of Earth and Mineral Sciences, The Pennsylvania State University

Kersters I, Vanvooren L, Huys G, Janssen P, Kersters K, Verstraete W (1995) Influence of temperature and process technology on the occurrence of aeromonas species and hygienic indicator organisms in drinking-water production plants. Microb Ecol 30(2):203–218

Kuan DT, Sawchuk AA, Strand TC, Chavel P (1985) Adaptive noise smoothing filter for images with signal-dependent noise. IEEE Trans Pattern Anal Mach Intell PAMI-7(2):165–177

Massoud MA (2012) Assessment of water quality along a recreational section of the Damour River in Lebanon using the water quality index. Environ Monit Assess 184(7):4151–4160

Matgen P, Schumann G, Henry JB, Hoffmann L, Pfister L (2007) Integration of SAR-derived river inundation areas, high-precision topographic data and a river flow model toward near real-time flood management. Int J Appl Earth Obs Geoinf 9(3):247–263

Osode AN, Okoh AI (2010) Survival of free-living and plankton-associated Escherichia coli in the final effluents of a waste water treatment facility in a peri-urban community of the Eastern Cape Province of South Africa. Afr J Microbiol Res 4(13):1424–1432

Williams RJ, Boorman DB (2012) Modelling in-stream temperature and dissolved oxygen at sub-daily time steps: an application to the River Kennet, UK. Sci Total Environ 423:104–110

Chapter 3
Flooding Identification by Vegetation Index

In this chapter, flooding identification by vegetation index was conducted by using SAR images and multispectral images in order to obtain the flooding period of the study area. The results of this flooding identification method would be used in the second part to apply to waterborne diseases caused by flooding disaster. This flooding identification method is simple to implement and works without complicated processes. However, since the validated data is limited, only eight districts were included in this flooding identification method.

3.1 Flooding Identification from Multispectral Images

Within the visible (VIS) and infrared (IR) spectral bands, water can be characterized by a high reflectance in the blue wavelength, which rapidly diminishes in the visible wavelengths to become very weak in near infrared (NIR). Based on this spectral signature, several indices were developed to delineate water bodies using ratios of spectral bands; the most frequent wavelengths used were NIR, green (G) and middle infrared (MIR). These spectral indices aim to maximize the difference of reflectance values between the object being studied (i.e., water bodies) and other surfaces. We can summarize the indices encountered in scientific literature and indicates how they are calculated and in which context they were originally used (Tran et al. 2010; as shown in Table 3.1). The wide use of spectral indices for water detection compose of:

(1) NDWI (normalized difference water index) is derived from normalized difference vegetation index (NDVI), adapted to the delineation of water bodies with the use of reflectance in the green wavelength (McFeeters 1996). It corresponds to green and near-infared bands on multispectral images. NDWI can be effectively used to reduce the effects of vegetation and highlight the information of water bodies.

© Higher Education Press and Springer Nature Singapore Pte Ltd. 2021
C. Cao et al., *Environmental Remote Sensing in Flooding Areas*,
https://doi.org/10.1007/978-981-15-8202-8_3

Table 3.1 Spectral indices used for water detection (Tran et al. 2010)

Name	Calculation	Context	Reference
NIR	–	Water detection	Work and Gilmer (1976), White (1978), Rundquist et al. (1987)
NDVI (normalized difference vegetation index)	$NDVI = \frac{NIR-R}{NIR+R}$	The index NDVI was originally used to assess the biomass and vegetation primary production. Nevertheless, it can be of interest to detect water as NDVI shows positive values for vegetation, values close to zero for bare soil and negative values for water	Rouse et al. (1973), Huete et al. (1997)
EVI (enhanced vegetation index)	$EVI = 2.5\frac{NIR-R}{NIR+6R-7.5B+1}$	EVI is derived from NDVI. EVI tends to limit the aerosols effects and minimize soil effects	Huete et al. (2002)
NDWI (normalized difference water index)	$NDWI = \frac{G-NIR}{G+NIR}$	NDWI is derived from NDVI, adapted to the delimitation of water bodies with the use of reflectance in the green wavelength	McFeeters (1996)
NDII (normalized difference infrared index); NDWI (normalized difference water index); NDMI (normalized difference moisture index); LSWI (land surface water index)	$NDII = NDWI = NDMI = LSWI = \frac{NIR-MIR}{NIR+MIR}$	Initially, Hardisky et al. (1983) showed a correlation between NDII and canopy water content. The NDWI (Gao 1996) and the NDMI (Wilson and Sader 2002) were used to detect leaves water content and to assess soil moisture. The LSWI (Xiao et al. 2005) is used to detect soil moisture from MODIS data	Hardisky et al. (1983), Gao (1996), Wilson and Sader (2002), Xiao et al. (2005)

(continued)

Table 3.1 (continued)

Name	Calculation	Context	Reference
MNDWI (modified normalized difference water index); INH (normalized humidity index); NDPI (normalized difference pond index)	MNDWI = −INH = −NDPI = $\frac{G-NIR}{G+NIR}$	The MNDWI is derived from the NDWI defined by McFeeters (1996) by the use of middle infrared instead of near infrared (Xu 2006). Water bodies are better delineated by a more efficient discrimination between open surface water and dry surfaces. The threshold of discrimination is located around 0. The INH was used by Clandillon et al. (1995) in order to detect humidity in wetland environment. The NDPI was used for detection of small ponds and streams semi-arid areas (Lacaux et al. 2007)	Xu (2006), Clandillon et al. (1995), Lacaux et al. (2007)

(2) NDVI was originally used to assess the biomass and vegetation primary production. Nevertheless, it can be of interest to detect water as NDVI shows positive values for vegetation, values close to zero for bare soil and negative values for water (Huete et al. 1997). Because the water bodies in the NDVI index are generally negative or close to 0, it can be used to map water bodies by using the threshold segmentation approach (Brakenridge and Anderson 2006).

(3) NDII (normalized difference infrared index) shows a correlation between NDII and canopy water content.

(4) MNDWI (modified normalised difference water index) is derived from the NDWI and defined by the use of middle infrared instead of near infrared (Xu 2006). Water bodies are better delineated by a more efficient discrimination between open surface water and dry surfaces. The threshold of discrimination is located around 0.

With the characteristic of THEOS and HJ-1A/B multispectral imagery used in this study, with no MIR band, a NDWI(McFeeters 1996) was selected to extract water from a series of THEOS multispectral imagery. NDWI has been very widely used for remote sensing to delineate open water features for many years and are defined as:

$$\mathrm{NDWI} = \frac{\rho_{\mathrm{Green}} - \rho_{\mathrm{NIR}}}{\rho_{\mathrm{Green}} + \rho_{\mathrm{NIR}}} \tag{3.1}$$

where ρ_{Green} and ρ_{NIR} are the reflectance of the green and NIR bands respectively.

NDWI is designed to (i) maximize the reflectance of a water body by using green wavelengths, (ii) minimize the low reflectance in NIR of water bodies, and (iii) take advantage of the high reflectance in NIR of vegetation and soil features (Li et al. 2013).

Threshold segmentation is a key step in extracting water bodies from the background (Lu et al. 2011). The threshold values for McFeeters's and Xu's NDWIs were set to zero (Li et al. 2013; Lu et al. 2011), but threshold adjustment in individual situations can achieve a more accurate delineation of water bodies (Ji et al. 2009). Hence, dynamic or varied thresholds are needed when different regions or different phases of remote sensing data are employed to detect water body information (Du et al. 2012). The maximum between-class variance method (the Otsu method) is one such dynamic threshold method (Otsu 1979). In this study, the Otsu method was used to determine the threshold for separating water bodies from the background features (Li et al. 2013; Lu et al. 2011). The Otsu method selects the threshold by employing the rule of optimum between class variance of water body features and the background features such as vegetation, soil, etc. When a number of the water body features are wrongly classified to a background feature, or a number of the background features are mistakenly classified to a water body feature, the between-class variance decreases. This implies that the greater the variance, the more different the background features and the water body features. Consequently, maximizing the variance between water body features and background features can minimizes the probability of misclassification. Otherwise, a spectral water index model is more appropriate for enhancing and classifying water body feathers when the NDWI image has high between class variance.

The validity of NDWIs can be evaluated from an image segmentation perspective. The contrast method (CM) (Xu 2006) was used to assess the adequacy of water body information extraction by using the NDWI. The method compares the digital number (DN) obtained from the distribution regions of water body and its background features in the same index. The contrast value (CV) between water bodies and background features can be obtained by

$$\mathrm{CV} = |M_{\mathrm{W}} - M_{\mathrm{nW}}| \tag{3.2}$$

where M_{W} and M_{nW} are the mean pixel values of the water class and the non-water class respectively. In addition, the extracted flood boundaries were rechecked through a visual interpretation approach with expert flood knowledge to evaluate the water body mapping results (Oguro et al. 2003). In this study, we employ a data set of THEOS and HJ-1A/B multi-spectral imagery having four bands of red, green,

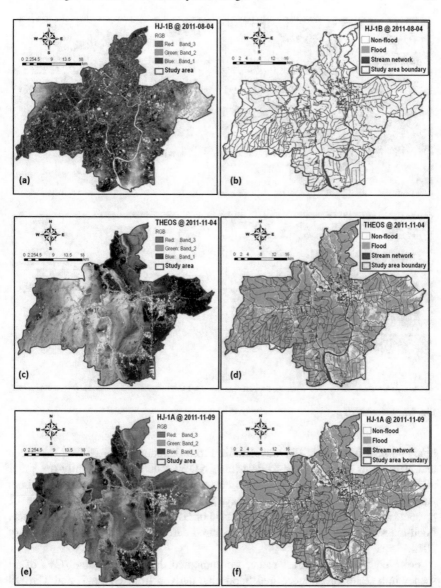

Fig. 3.1 A dataset of multi-spectral satellite imagery used in the study include HJ-1A/B with 30-m resolution on 4 August 2011 (**a**), 9 November 2011 (**e**), and 24 January 2012 (**i**), and THEOS with 15-m resolution on 4 November 2011 (**c**) and 1 December 2011 (**g**). A series of classified flood map derived from HJ-1A/B and THEOS imageries on 4 August 2011 (**b**), 4 November 2011 (**d**), 9 November 2011 (**f**), 1 December 2011 (**h**), and 24 January 2012 (**j**)

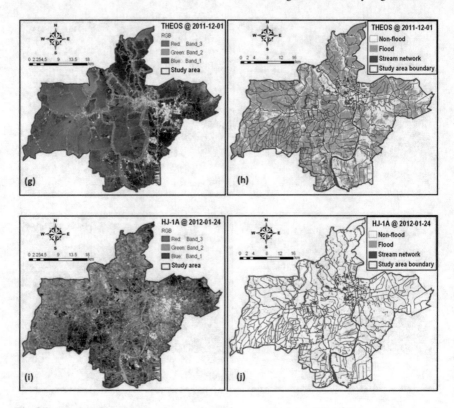

Fig. 3.1 (continued)

blue and near-infrared (NIR) wavelength. From the equation of NDWI, the green and NIR bands were utilized to determine NDWI values, subsequently the water and non-water areas were classified by a thresholding and visualization method. Analysis of the eight districts in Ayutthaya Province applied this method. Figure 3.1 presents a series of the multi-spectral image of HJ-1A/B and THEOS (left) with its flood-classified map (right) sorted by acquired date from August 2011 to January 2012.

From equation of contrast value, we computed the contrast value (CV) of all scenes in a series of the classified flood map derived from HJ-1A/B and THEOS imageries to evaluate the performance of NDWI classified enhancement. The range of CV during 0.42–0.55, as shown in Table 3.2 indicates that the perspective of image segmentation is appropriate to the water body classified images (Li et al. 2013).

	Satellite	Date	Contrast value
Table 3.2 The contrast values from the threshold segmentation of multi-spectral images of HJ-1A/B and THEOS indicate that the resulting flood classified maps are efficient	HJ-1B	4-Aug-2011	0.47
	THEOS	4-Nov-2011	0.47
	HJ-1A	9-Nov-2011	0.42
	THEOS	1-Dec-2011	0.55
	HJ-1A	24-Jan-2012	0.51

3.2 Flooding Identification from SAR Images

Based on SAR data, the speckle noise in the Radars at images was reduced by utilizing preprocessing. To derive flood extent maps from airborne or satellite imagery, prior to image processing, a filter such as Kuan filters should be applied to SAR imagery to remove most speckle, i.e. random image noise obstructing features of interest. Then, many different image-processing techniques may be applied to a satellite image. However, it is well-known that no single method can be considered appropriate for all images, nor are all methods equally suitable for a particular type of image. The most common procedures are visual interpretation, histogram threshold, active contour, and image texture variance. The advantages and disadvantages are shown in Table 3.3.

Visual interpretation approach: a flooded area is mapped by visually digitizing the flood boundaries. A skillful delineation of flood shorelines by visual interpretation requires expert flood knowledge. If breaks of slope between the floodplain and adjacent hill slopes are clearly visible in the topography, they should be used to constrain the delineated flood extent to the valley floor area.

Histogram thresholding is a simple but widely used and efficient method to generate binary maps from images. An optimal grey level threshold can be found using the Otsu method. The method applies a criterion measure to evaluate the between-class variance (i.e. separability) of a threshold at a given level computed from a normalized image histogram of gray levels.

The active contour method is based on a dynamic curvilinear contour that searches the edge image space until it settles upon image region boundaries. This is achieved by an energy function attracted to edge points. The contour is usually represented as a series of nodes linked by straight line segments (Horritt and Bates 2001). The statistical snake is formulated as an energy minimization. The total energy is minimized if the contour encloses a large area of good pixels, and in this respect the model behaves as a region growing algorithm.

Image texture can be modeled as a grey level function using simple statistical methods on the image histogram. Widely used algorithms rely on statistical properties of a neighbourhood of pixels, which are computed for each pixel using a moving window. The image texture variance and mean Euclidean distance (Irons and Petersen 1981) can be found in most commercial remote sensing software packages. However, histogram thresholding (Hostache et al. 2009) and visual interpretation (Oberstadler

Table 3.3 Advantages and disadvantages of commonly used image processing techniques to obtain flood area from SAR images (Di Baldassarre et al. 2011)

	Visual interpretation	Histogram thresholding	Texture based	Active contour modeling/Region growing
Strength	Easy to perform in case of a skilled and experienced operator with knowledge of flood processes	Easy and quick to apply Objective method	Takes account of the SAR textural variation Based on statistics Mimics human interpretation as it takes account of tonal differences	Image statistics based Usually provides good classification results Easy to define seed region (e.g. on the river channel) If integrated with land elevation constraints results are improved by mimicking inundation processes
Limitation	Very subjective Difficult to implement over many images May be difficult for images that show complex flood paths	No flexibility Optimized threshold might not be the most appropriate Works only well if image is relatively little distorted	Difficult to choose correct window size and appropriate texture measure After application still requires threshold value to obtain flood area classification	Requires several parameters to fine-tune Slow on large image domains Difficult to choose correct tolerance criterion May miss separated patches of dry or flooded land
Level of complexity	Low to high (may have varying degrees of complexity)	Very low	Moderate	Moderate to high
Computational efficiency	Relatively low	Very low	Moderate	Moderate (strongly depending on domain size)
Level of automation	Hardly possible	Full	Full	Relatively high
Consistency	0.9	0.8	0.6	0.7

et al. 1997) are very popular methods to delineate flood areas, and therefore we will employ a combination of these two methods in this study (Rakwatin et al. 2013). An appropriate threshold is chosen by visual inspection of the image histogram (Matgen et al. 2007). Threshold values are manually selected for each image individually using visual interpretation to label areas as flood or dry. To avoid misidentification due to radar backscattering of asphalt roads and permanent water bodies, which may be similar to the inundated area (Badji and Dautrebande 1997), we utilized GIS layers including road and hydrographic features to overlay the extracted water map(Waisurasingha et al. 2008).

For this study, the flooded area was extracted from SAR imagery. SAR has significant advantages for the detection of water bodies and can penetrate clouds (Schumann et al. 2009). Radarsat is in operational use for flood monitoring in many countries. It has been shown to accurately assess and clarify inundated areas. Moreover, its ability to penetrate clouds is very important for monitoring floods during the rainy season in monsoon countries (Hoque et al. 2011). Radarsat-2 is the second in a series of Canadian spaceborne SAR satellites that provides several improvements over Radarsat-1, such as additional beam modes, higher resolution, multi-polarization, and more-frequent revisits. Our time series of Radarsat-2 scenes with 50-m resolution in ScanSAR narrow mode acquired from September to December 2011, was manually ortho rectified with topographic base maps, and the nearest-neighbor method was used to preserve original values in the re-sampling process. The Universal Transverse Mercatorzone 47 was defined as the image-to-map projection. The acceptable threshold of the Root Mean Square (RMS) error was set to one pixel due to limited human resources and time constraints (Rakwatin et al. 2013), and this error was considered as a buffer applied to the extraction process for areas of water bodies. The most widely used adaptive filters based on the spatial domain to reduce the speckle noise in the SAR images include the Lee, Frost, Enfrost, Kuan, Median, and Gamma filters (Matgen et al. 2007). Using trial and error with visualization, we applied the 5×5 Kuan filter to reduce the speckle noise for our Radarsat-2 images (Gupta and Gupta 2007; Matgen et al. 2007). Based on the criterion of minimum mean square error, the Kuan filter applies a spatial filter to each pixel that is replaced with a value calculated based on the local statistics and can reduce speckle while preserving edges by transforming the multiplicative noise model into an additive noise model (Kuan et al. 1985; Shi and Fung 1994).

Histogram thresholding (Hostache et al. 2009) and visual interpretation (Oberstadler et al. 1997) are popular methods for delineating flooded areas based on SAR data. We therefore employed a combination of these two methods in this study (Rakwatin et al. 2013) as shown in Fig. 3.2. An appropriate threshold was chosen by visual inspection of the image histogram (Matgen et al. 2007). Threshold values were manually selected for each image individually using visual interpretation to label areas as flooded or dry. To avoid misidentification due to radar backscattering from asphalt roads and permanent water bodies that may appear similar to inundated areas (Badji and Dautrebande 1997), we utilized GIS layers that included road and hydrographic features to overlay with the extracted water map (Waisurasingha et al. 2008). A flood series map derived from six Radarsat-2 scenes is presented in

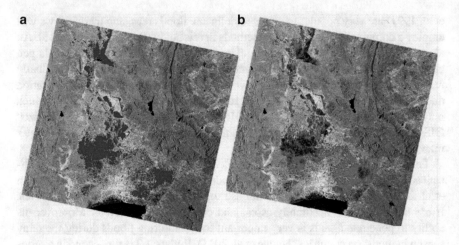

Fig. 3.2 Each Radarsat-2 imagery is delineated water body by using histogram thresholding (**a**) and visual interpretation (**b**) approach respectively

Fig. 3.3. The resulting flooded areas in six Radarsat-2 scenes show that flooding began approximately on 9 September (Fig. 3.3b) and continuously increased on 3 October (Fig. 3.3d) and 21 October (Fig. 3.3f) until its peak on 14 November (Fig. 3.3h). Subsequently, the flood gradually abated on 4 December (Fig. 3.3j) and 28 December (Fig. 3.3l).

3.3   Multi-temporal Remote Sensing Data for Flooding Identification

To efficiently monitor flood events, we need remote sensing data as much as possible. However, the quantity of remote sensing data from a single satellite is always limited by the revisiting time of each Earth Observation (EO) satellite and the negative influences of clouds. Multi-sensor and multi-temporal EO satellite imageries can provide us with more frequently updated flood information than using only one EO satellite data (Mcginnis and Rango 1975). In recent years using a multi-satellite method, including passive microwave land surface emissivities, along with active microwave, visible and near infrared bandwidth observations developed to estimate global inundated areas (Papa et al. 2007; Prigent et al. 2007) and examine the spatial and temporal variations of the 1993–2000 monthly inundations in the Ob River basin and their relation at the local scale with snow depth measurements and river run-off and discharge. Pappas et al. (2008) developed a multi-satellite method for employing passive and active microwaves along with visible and infrared observations to estimate monthly inundation extent at global scale. They demonstrated that the monthly

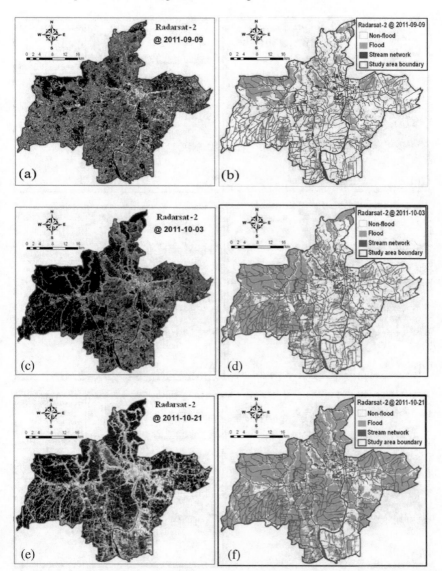

Fig. 3.3 A dataset of SAR imagery used in the study includes Radarsat-2 images with 50-m resolution on 9 September 2011 (**a**), 3 October 2011 (**c**), 21 October 2011 (**e**), 14 November 2011 (**g**), 4 December 2011 (**i**), and 8 December 2011 (**k**). Meanwhile, a series of classified flood map is derived from the Radarsat-2 imageries acquired on 9 September 2011 (**b**), 3 October 2011 (**d**), 21 October 2011 (**f**), 14 November 2011 (**h**), 4 December 2011 (**j**), and 28 December 2011 (**l**)

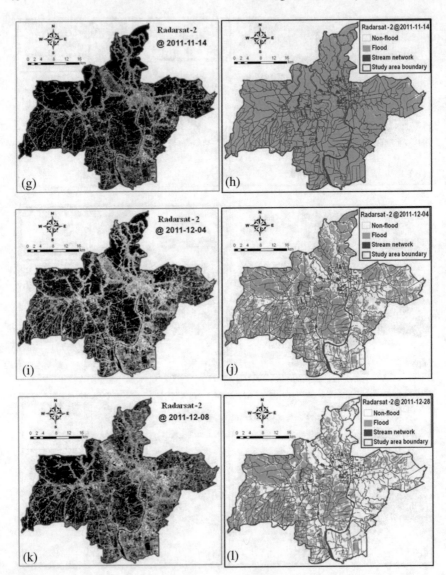

Fig. 3.3 (continued)

multi-satellite-derived inundation dataset brings a new useful tool for better understanding both the stream flow processes and the description of the snow-inundation runoff relations in data scarce areas like the remote Arctic river basins.

This study combined flood areas of the multi-sensor and multi-temporal remote sensing data (Brivio et al. 2002; Oberstadler et al. 1997) so as to delineate a flood map series (Brivio et al. 2002; Hostache et al. 2009; Rakwatin et al. 2013). The techniques for flood extraction of passive and active remote sensing data will be

Fig. 3.4 Using GPS vehicle-tracking devices by GISTDA to validate flood areas from multi-sensor and multi-temporal remote sensing data

utilized along with validating methods (Butenuth et al. 2011). The flood map series will be validated by ground truth data acquired from field surveys and flood related organizations (Rakwatin et al. 2013).

With the ability to penetrate clouds and a lack of weather condition restrictions, we employed the series of Radarsat-2 imagery as the main data for delineating the temporal changes of flood areas. Meanwhile, the multi-spectral datasets of HJ-1A/B and THEOS were used to fulfil the flood detection and increase the temporal resolution of flood monitoring.

The output flood maps were validated with ground data such as reports of flood-relief officers and broadcast news, which were collected by the Geo-Informatics and Space Technology Development Agency (GISTDA) (Rakwatin et al. 2013) to classify each pixel as 'flood' or 'non-flood'. Figure 3.4 provides the GPS vehicle-tracking devices composed of car, mobile GPS, laptop and tracking applications. The GPS vehicle-tracking devices were used for validating the resulting flood areas from the remote sensing method by tracking along roads and reviewing with the local people in the case of inaccessible areas. However, during the major flood disaster validating the classified flood areas quantitatively constituted an arduous task, because flood relief officers and reporters may be unable to access and report on the flood-affected areas and comprehensivestatistical data not yet collected.

Combining the flood classification results from the multi-temporal satellite data of Radarsat-2, HJ-1A/B and THEOS, we can observe the apparent, dynamic change of flooding. From Figs. 3.1 and 3.5, flooding in the study area started approximately in the beginning of September 2011, then the flood area gradually increased from September until its peak in November 2011, when flood water covered almost sections of the study area. The flood level gradually decreased in the beginning of December 2011 until it completely dried in the end of January 2012.

The change of flood areas at various time by monitoring of multi-temporal satellite included Radarsat-2, HJ-1A/B and THEOS imageries

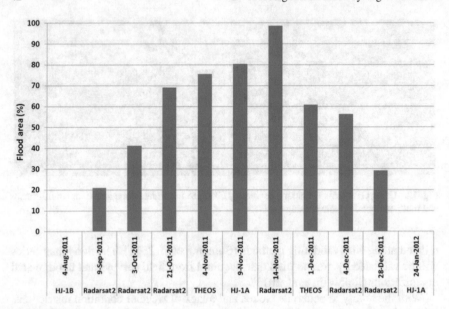

Fig. 3.5 Percentage of flood area from the multi-temporal remote sensing data at various times

3.4 Summary

Based on optical imagery, several water spectral indices were developed to delineate water bodies by using ratios of spectral bands. With the characteristic of THEOS and HJ-1/AB multispectral imagery, NDWI, a widely used, proven index for water detection, was employed to delineate water bodies during flooding from the multispectral imageries in this study. With the significant advantages of SAR data for the detection of water bodies and penetrating clouds, a series of Radarsat-2 imageries were used to extract water features in this study by using a combination of two popular SAR methods, which are histogram thresholding and visual interpretation. A series of water body maps from multi-sensor and multi-temporal remote sensing data was overlaid with GIS layers, such as road and hydrographic features, to define flood areas. The output flood map was validated with ground data to classify each pixel as 'flood' or 'non-flood'. The resulting flooded areas show that the flooding of our study area in 2011 began approximately the beginning of September and continuously increased in October until its peak in the middle of November. Subsequently, the flood gradually abated from the beginning of December 2011 until it completely dried in January 2012.

References

Badji M, Dautrebande S (1997) Characterization of flood inundated areas and delineation of poor drainage soil using ERS-1 SAR imagery. Hydrol Process 11(10):1441–1450

Baldassarre D, Schumann G, Brandimarte L, Bates P (2011) Timely low resolution SAR imagery to support floodplain modelling: a case study review. Surveys in Geophysics 32:255–269

Brakenridge R, Anderson E (2006) Modis-based flood detection, mapping and measurement: the potential for operational hydrological applications. Transbound Floods: Reduc Risks Through Flood Manage 72:1–12

Brivio PA, Colombo R, Maggi M, Tomasoni R (2002) Integration of remote sensing data and GIS for accurate mapping of flooded areas. Int J Remote Sen 23(3):429–441

Butenuth M, Frey D, Nielsen AA, Skriver H (2011) Infrastructure assessment for disaster management using multi-sensor and multi-temporal remote sensing imagery. Int J Remote Sen 32(23):8575–8594

Clandillon S, De Fraipont P, Yesou H (1995) Assessment of the future SPOT 4 MIR for wetland monitoring and soil moisture analysis. A simulation case over the Grand Ried d'Alsace (France). In: European Symposium on Remote Sensing II.

Du ZQ, Bin LH, Ling F, Li WB, Tian WD, Wang HL, Gui YM, Sun BY, Zhang XM (2012) Estimating surface water area changes using time-series Landsat data in the Qingjiang River Basin, China. J Appl Remote Sen 6

Gao BC (1996) NDWI—a normalized difference water index for remote sensing of vegetation liquid water from space. Remote Sen Environ 58:257–266

Gupta KK, Gupta R (2007) Despeckle and geographical feature extraction in SAR images by wavelet transform. Isprs J Photogram Remote Sen 62(6):473–484

Hardisky MA, Klemas V, Smart RM (1983) The influences of soil salinity, growth form, and leaf moisture on the spectral reflectance of Spartina alterniflora canopies. Photogrammetric Engineering and Remote Sensing 49:77-83

Hoque R, Nakayama D, Matsuyama H, Matsumoto J (2011) Flood monitoring, mapping and assessing capabilities using RADARSAT remote sensing, GIS and ground data for Bangladesh. Nat Hazards 57(2):525–548

Horritt MS, Bates PD (2001) Effects of spatial resolution on a raster based model of flood flow. J Hydrology 253(1–4):239–249

Hostache R, Matgen P, Schumann G, Puech C, Hoffmann L, Pfister L (2009) Water level estimation and reduction of hydraulic model calibration uncertainties using satellite SAR images of floods. IEEE Trans Geosci Remote Sen 47(2):431–441

Huete AR, Liu HQ, Batchily K, vanLeeuwen W (1997) A comparison of vegetation indices global set of TM images for EOS-MODIS. Remote Sen Environ 59(3):440–451

Huete A, Didan K, Miura T, Rodriguez EP, Gao X, Ferreira LG (2002) Overview of the radiometric and biophysical performance of the MODIS vegetation indices. Remote Sen Environ 83:195-213

Irons JR, Petersen GW (1981) Texture transforms of remote sensing data. Remote Sen Environ 11(5):359–370

Ji L, Zhang L, Wylie B (2009) Analysis of Dynamic Thresholds for the Normalized Difference Water Index. Photogram Eng Remote Sen 75(11):1307–1317

Kuan DT, Sawchuk AA, Strand TC, Chavel P (1985) Adaptive noise smoothing filter for images with signal-dependent noise. IEEE Trans Pattern Anal Mach Intell 7(2):165–177

Lacaux JP, Tourre YM, Vignolles C, Ndione JA, Lafaye M (2007) Classification of ponds from high-spatial resolution remote sensing: application to rift valley fever epidemics in Senegal. Remote Sen Environ 106:66-74

Li WB, Du ZQ, Ling F, Zhou DB, Wang HL, Gui YM, Sun BY, Zhang XM (2013) A comparison of land surface water mapping using the normalized difference water index from TM, ETM plus and ALI. Remote Sen 5(11):5530–5549

Lu SL, Wu BF, Yan NN, Wang H (2011) Water body mapping method with HJ-1A/B satellite imagery. Int J Appl Earth Obs Geoinf 13(3):428–434

Matgen P, Schumann G, Henry JB, Hoffmann L, Pfister L (2007) Integration of SAR-derived river inundation areas, high-precision topographic data and a river flow model toward near real-time flood management. Int J Appl Earth Obs Geoinf 9(3):247–263

McFeeters SK (1996) The use of the normalized difference water index (NDWI) in the delineation of open water features. Int J Remote Sen 17(7):1425–1432

Mcginnis DF, Rango A (1975) Earth resources satellite systems for flood monitoring. Geophys Res Lett 2(4):132–135

Oberstadler R, Honsch H, Huth D (1997) Assessment of the mapping capabilities of ERS-1 SAR data for flood mapping: a case study in Germany. Hydrol Process 11(10):1415–1425

Oguro Y, Takeuchi S, Suga Y, Tsuchiya K (2003) Higher resolution images for visible and near infrared Bands of Landsat-7 ETM + by using panchromatic Band. Calibration, Characterization of Satellite Sensors, Physical Parameters Derived from Satellite Data 32(11):2269–2274

Otsu N (1979) Threshold selection method from gray-level histograms. IEEE Trans Syst Man Cybern 9(1):62–66

Papa F, Prigent C, Rossow WB (2007) Ob' River flood inundations from satellite observations: A relationship with winter snow parameters and river runoff. J Geophys Res-Atmos 112(D18)

Pappas G, Papadimitriou P, Siozopoulou V, Christou L, Akritidis N (2008) The globalization of leptospirosis: worldwide incidence trends. Int J Infect Dis 12(4):351-357

Prigent C, Papa F, Aires F, Rossow WB, Matthews E (2007) Global inundation dynamics inferred from multiple satellite observations, 1993–2000. J Geophys Res-Atmos 112(D12)

Rakwatin P, Sansena T, Marjang N, Rungsipanich A (2013) Using multi-temporal remote-sensing data to estimate 2011 flood area and volume over Chao Phraya River basin Thailand. Remote Sen Lett 4(3):243–250

Rouse JW, Haas RH, Schell JA, Deering DW (1973) Monitoring vegetation systems in the great plains with ERTS. 3rd ERTS Symposium 48-62

Rundquist DC, Lawson MP, Queen LP, Ceverny RS (1987) The relationship between summer-season rainfall events and lake-surface area. Water Resources Bulletin 23:493-508

Schumann G, Bates PD, Horritt MS, Matgen P, Pappenberger F (2009) Progress in Integration of Remote Sensing-Derived Flood Extent and Stage Data and Hydraulic Models. Rev Geophys 47

Shi ZG, Fung KB (1994) A comparison of digital speckle filters. In: IGARSS '94—1994 international geoscience and remote sensing symposium, vol 1–4, pp 2129–2133

Tran A, Goutard F, Chamaille L, Baghdadi N, Lo Seen D (2010) Remote sensing and avian influenza: a review of image processing methods for extracting key variables affecting avian influenza virus survival in water from Earth Observation satellites. Int J Appl Earth Obs Geoinf 12(1):1–8

Waisurasingha C, Aniya M, Hirano A, Sommut W (2008) Use of RADARSAT-1 data and a digital elevation model to assess flood damage and improve rice production in the lower part of the Chi River Basin Thailand. Int J Remote Sen 29(20):5837–5850

White ME (1978) Reservoir surface area from Landsat imagery. Photogrammetric Engineering and Remote Sensing 44:1421-1426

Wilson EH, Sader SA (2002) Detection of forest harvest type using multiple dates of Landsat TM imagery. Remote Sen Environ 80:385-396

Work EA, Gilmer DS (1976) Utilization of satellite data for inventorying prairie ponds and lakes. Photogrammetric Engineering and Remote Sensing 42:685-694

Xiao XM, Boles S, Liu JY, Zhuang DF, Frolking S, Li CS, Salas W, Moore B (2005) Mapping paddy rice agriculture in southern China using multi-temporal MODIS images. Remote Sen Environ 95:480-492

Xu HQ (2006) Modification of normalised difference water index (NDWI) to enhance open water features in remotely sensed imagery. Int J Remote Sen 27(14):3025-3033

Chapter 4
Flooding Identification by Support Vector Machine

4.1 Support Vector Machine

Support Vector Machine (SVM) is an advanced algorithm and widely used in many applications. In this chapter, Support Vector Machines (SVMs) are described. The purpose of SVM is to find a hyperplane that separates the data (Boswell 2002; Fletcher 2009; Huang et al. 2002; Lindsay 2003; Theodoridis and Koutroumbas 2008; Theodoridis et al. 2010; Venturad 2009; Watanachaturaporn et al. 2005). The following would briefly explain the concept of SVM.

1. SVM in Linear Separate Binary Classification

In linear separate binary classification the hyperplane expresses as Fig. 4.1 and Eq. (4.1).

$$H = \mathbf{w}^{\mathrm{T}} x + w_0 = 0 \tag{4.1}$$

$\mathbf{x}^i = \left\{ x_1^i, x_2^i, \ldots, x_n^i \right\}$ is the n-dimensional vectors for the ith example in the dataset. These training data belong to the class labeled as y_i. Hyperplane direction is symbolized by $\mathbf{w}$ and its exact position in space determines by w_0. The importance of the concept of the SVM classifier design is a maximizing margin, which is the distance between the two parallel hyperplanes and follow as an Eqs. (4.2) and (4.3).

$$\mathbf{w}^{\mathrm{T}} x + w_0 = +1, \text{ for } H_1 \tag{4.2}$$

$$\mathbf{w}^{\mathrm{T}} x + w_0 = -1, \text{ for } H_1 \tag{4.3}$$

The Euclidean distance of any sample points that lie on either of the two hyperplanes from the classifier hyperplane is equal to $1/\|\mathbf{w}\|$, where $\|\mathbf{w}\|$ is the Euclidean norm. Therefore, obtaining a margin of $1/\|\mathbf{w}\| + 1/\|\mathbf{w}\| = 2/\|\mathbf{w}\|$, and aiming to get

© Higher Education Press and Springer Nature Singapore Pte Ltd. 2021
C. Cao et al., *Environmental Remote Sensing in Flooding Areas*,
https://doi.org/10.1007/978-981-15-8202-8_4

Fig. 4.1 Hyperplane
separates the data entirely

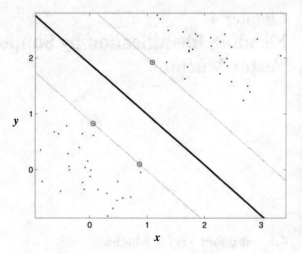

the variables w and w_0 is important so that the training data can be described by:

$$w^Tx + w_0 \geq 1, \forall x \in y_1 \tag{4.4}$$

$$w^Tx + w_0 \leq -1, \forall x \in y_2 \tag{4.5}$$

Equations (4.4) and (4.5) can be combined:

$$y_i(w^Tx_i + w_0) \geq 1, i = 1, \cdots, N \tag{4.6}$$

The maximum margin is needed in constructing an SVM. The classifier, the minimizing $\|w\|$, is equivalent to the technique of minimizing $\frac{1}{2}\|w\|^2$ taken into consideration. Compute the parameters w and w_0 of the hyperplane:

$$\begin{aligned} \text{minimize} \quad & \frac{1}{2}\|w\|^2 \\ \text{subject to} \quad & y_i(w^Tx_i + w_0) \geq 1, \forall_i \end{aligned} \tag{4.7}$$

Practically applied, this minimizing process can be done by Quadratic Programming (QP) optimization. To accomplish this, a Lagrange multiplier (Fletcher 2009; Theodoridis and Koutroumbas 2008), α where $\alpha_i \geq 0, \forall_i$, needs to be allocated. The problem eventually transforms into:

$$L_p = L(w, w_0, \alpha) = \frac{1}{2}w^Tw - \sum_{i=1}^{N} \alpha_i[y_i(w^Tx_i + w_0) - 1] \tag{4.8}$$

The α in the Eq. (4.8) should be maximized resulting in w and w_0 minimized. To find w and w_0, differentiate $L(w, w_0, \alpha)$ with respect to w and w_0 setting the derivatives to zero:

$$\frac{\partial L_p}{\partial w} = \frac{\partial L(w, w_0, \alpha)}{\partial w} \rightarrow w = \sum_{i=1}^{N} \alpha_i y_i x_i \tag{4.9}$$

$$\frac{\partial L_p}{\partial w_0} = \frac{\partial L(w, w_0, \alpha)}{\partial w_0} \rightarrow \sum_{i=1}^{N} \alpha_i y_i = 0 \tag{4.10}$$

Substituting Eqs. (4.9) and (4.10) into (4.8) to make the equation in terms of α, which is needed to maximize:

$$L_D = \sum_{i=1}^{N} \alpha_i - \frac{1}{2} \sum_{i,j} \alpha_i \alpha_j y_i y_j x_i^T x_j$$

$$\text{subject to} \quad \alpha_i \geq 0, \quad \forall_i$$

$$\sum_{i=1}^{N} \alpha_i y_i = 0 \tag{4.11}$$

$$L_D = \sum_{i=1}^{N} \alpha_i - \frac{1}{2} \sum_{ij} \alpha_i H_{ij} \alpha_j, \quad H_{ij} = y_i y_j x_i x_j$$

$$L_D = \sum_{i=1}^{N} \alpha_i - \frac{1}{2} \alpha^T H \alpha$$

$$\text{subject to} \quad \alpha_i \geq 0, \quad \forall_i$$

$$\sum_{i=1}^{N} \alpha_i y_i = 0 \tag{4.12}$$

L_D is the dual form of the primary L_p. Now, moving from minimizing L_p to maximizing L_D, one must find:

$$\max_{\alpha} \left[\sum_{i=1}^{N} \alpha_i - \frac{1}{2} \alpha^T H \alpha \right]$$

$$\text{subject to} \quad \alpha_i \geq 0, \quad \forall_i$$

$$\sum_{i=1}^{N} \alpha_i y_i = 0 \tag{4.13}$$

By implementing of the quadratic optimization problem, $\boldsymbol{\alpha}$ will return and according to the Eq. (4.9), the variable $\boldsymbol{w}$ can be found. Another variable to build the SVM classifier is w_0. Support vectors (x_s) are determined by the data points that satisfy Eq. (4.10), which can be written in the form:

$$y_s(x_s \cdot \boldsymbol{w} + w_0) = 1$$

Substituting in Eq. (4.9)

$$y_s\left(\sum_{m \in S} \alpha_m y_m x_m \cdot x_s + w_0\right) = 1 \tag{4.14}$$

S denotes the set of indices of the support vectors. S is determined by finding the indices i where $\alpha_i > 0$. Using y_s, multiply both sides of Eq. (4.14)

$$y_s^2\left(\sum_{m \in S} \alpha_m y_m x_m \cdot x_s + w_0\right) = y_s$$

According to Eqs. (4.4) and (4.5), $y_s^2 = 1$

$$w_0 = y_s - \sum_{m \in S} \alpha_m y_m x_m \cdot x_s$$

Therefore, the average over all of the support vectors in S becomes:

$$w_0 = y_s - \sum_{m \in S} \alpha_m y_m x_m \cdot x_s \tag{4.15}$$

Now, $\boldsymbol{w}$ and w_0 are available for creating the hyperplane of the SVM by the following Eqs. (4.1), (4.2) and (4.3).

2.   SVM in Binary Classification for Data Not Fully Linearly Separated

In event of the binary classification of data that is not entirely linearly separated (Fig. 4.2). The Eqs. (4.4) and (4.5) have been modified to allow for the misclassified points. This is performed by implementing a positive slack variable $\xi_i, i = 1, \cdots, N$

$$\boldsymbol{w}^T x + w_0 \geq +1 - \xi_i \quad \text{for } y_i = +1 \tag{4.16}$$

$$\boldsymbol{w}^T x + w_0 \leq -1 + \xi_i \quad \text{for } y_i = -1 \tag{4.17}$$

Combine Eqs. (4.16) and (4.17):

$$y_i[\boldsymbol{w}^T x + w_0] \geq 1 - \xi_i, \quad \xi_i \geq 0 \ \forall_i \tag{4.18}$$

Fig. 4.2 Hyperplanes do not fully separate the data

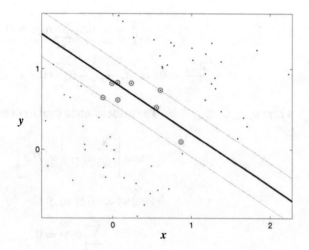

$\xi_i = 0$: The training feature vectors fall outside the margin and are correctly classified.

$0 < \xi_i \leq 1$: The training feature vectors fall inside the margin and are properly classified.

$\xi_i > 1$: The vectors are misclassified.

The goal of this case is the same in the previous cause, which is to have large margins but at the same time to keep the number of points with $\xi > 0$ as small as possible. Therefore, the objective function in Eq. (4.7) becomes:

$$\text{minimize} \quad \frac{1}{2}||w||^2 + C \sum_{i=1}^{N} I(\xi_i)$$

$$\text{subject to} \quad y_i(w^T x_i + w_0) - 1 + \xi_i \geq 0, \quad \forall_i \qquad (4.19)$$

The parameter C is a positive constant that controls the relative influence of the slack variable penalty and the size of the margin. Implement a Lagrangian as done before to minimize w, w_0, ξ_i and maximize with respect to α given by:

$$L_p = \frac{1}{2}||w||^2 + C \sum_{i=1}^{N} \xi_i - \sum_{i=1}^{N} \mu_i \xi_i - \sum_{i=1}^{N} \alpha_i \left[y_i(w^T x_i + w_0) - 1 + \xi_i \right] \qquad (4.20)$$

Differentiate with respect to w, w_0 and ξ_i set the derivatives to zero:

$$\frac{\partial L_p}{\partial w} = 0 \quad \text{or} \quad w = \sum_{i=1}^{N} \alpha_i y_i x_i \qquad (4.21)$$

$$\frac{\partial L_p}{\partial w_0} = 0 \quad \text{or} \quad \sum_{i=1}^{N} \alpha_i y_i = 0 \tag{4.22}$$

$$\frac{\partial L_p}{\partial \xi_i} = 0 \quad \text{or} \quad C - \mu_i - \alpha_i = 0, \quad i = 1, \cdots, N \tag{4.23}$$

where $\mu_i \geq 0, \alpha_i \geq 0, \forall_i$. The association to dual representation becomes:

$$\max_{\alpha} \left[\sum_{i=1}^{N} \alpha_i - \frac{1}{2} \alpha^T H \alpha \right]$$

$$\text{subject to} \quad 0 \leq \alpha_i \leq C, \quad \forall_i$$

$$\sum_{i=1}^{N} \alpha_i y_i = 0 \tag{4.24}$$

As with case 1, w_0 is then calculated.

3. Nonlinear Support Vector Machines

Nonlinear support vector machines, as in the previous linearly separable case, are designed SVM classifier which begin by building a matrix H from the dot product of input variables:

$$H_{ij} = y_i y_j k(x_i, x_j) = x_i \cdot x_j = x_i^T x_j \tag{4.25}$$

$k(x_i, x_j)$ is called a Kernel function. The linear kernel function is known as $k(x_i, x_j) = x_i^T x_j$. Thus, the functions can be moved into a higher dimensional space through a non-linear feature mapping function $x \rightarrow \phi(x)$, with $\phi(\cdot)$ denoting the inner product operation. Typical examples of kernels for classification are as follows:

Polynomial:

$$k(x_i, x_j) = (x_i \cdot x_j + a)^b$$

Radial Basis Function:

$$k(x_i, x_j) = \exp\left(-\frac{\|x_i - x_j\|^2}{2\sigma^2} \right)$$

Hyperbolic Tangent:

$$k(x_i, x_j) = \tanh(a x_i \cdot x_j - b)$$

where a and b are parameters defining the kernel's behavior. The steps of designing SVM classifiers consist of the steps in Table 4.1. In addition, Fig. 4.3 demonstrates the optimal hyperplane when applying different types of kernel functions.

According to Fig. 4.3 and Table 4.2, non-linear dataset separation can be accomplished by employing kernel functions that can map the data to a higher dimension.

4. Multiclass Support Vector Machine

In multiclass classification, two techniques are generally used. They normally maybe used with any classifier developed for the 2-class problem. The first technique is one-against-one. In this technique, the $C(C-1)/2$ binary classifier, where C is a number of classes, is trained and each classifier separates a pair of classes. The decision is made by majority vote. However, the disadvantage of this technique is that a large number of binary classifiers have to be trained. The second technique, which is attractive to use in this experiment, is the one-against-all technique. Each one of the c-classifiers is designed to separate one class from the rest. Therefore, it is required one design c linear classifiers:

$$w_i^T x + w_{0i}, \quad i = 1, \cdots, C$$

Table 4.1 Concluded steps of constructing SVM classifiers

Step no.	Process
1	Build H, where $H_{ij} = y_i y_j \phi(x_i) \cdot \phi(x_j)$, by choosing a kernel function and thus the mapping $x \to \phi(x)$
2	Determine the significant misclassifications that should be treated by choosing a suitable value of the parameter C
3	Find α so that: minimize $\sum_{i=1}^{N} \alpha_i - \frac{1}{2} \alpha^T H \alpha$ Subject to $0 \leq \alpha_i \leq C, \quad \forall_i$ $\sum_{i=1}^{N} \alpha_i y_i = 0$ This step is performed using a QP solver
4	Compute $w = \sum_{i=1}^{N} \alpha_i y_i \phi(x_i)$
5	Finding the indices such that $0 \leq \alpha_i \leq C$ determine the set of support vectors S
6	Compute $w_0 = \frac{1}{N_s} \sum_{s \in S} \left(y_s - \sum_{m \in S} \alpha_m y_m \phi(x_m) \cdot \phi(x_s) \right)$
7	For any given point x' is classified by evaluating $y' = sign(w \cdot \phi(x') + w_0)$

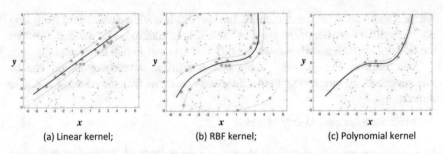

(a) Linear kernel;   (b) RBF kernel;   (c) Polynomial kernel

Fig. 4.3 Comparision of the optimal hyperplane by different kinds of kernel functions

Table 4.2 The parameter explanation of different kernel functions in Fig. 4.3

Parameter	Linear kernel	RBF kernel	Polynomial kernel
a	–	–	2
b	–	–	3
σ	–	2	–
C	2	5	5
Tolerance	0.001	0.01	0.01
No. support vector	26	26	8
Error (%)	7.33	1.33	0.67

The data from the rest of the classes result in negative outcomes. x is classified in w_i if $w^T x + w_{0,i} > w_j^T x + w_{0,j}, \quad \forall i \neq j$. Figure 4.4 shows an implementation of multiclass classification through the one-against-all technique. The example has four classes to separate by using a linear kernel function. For the first and second instance of finding the optimal hyperplane, class one and class two are considered to be the same category while class three and class four are set to be the same category, resulting in the hyperplane in Fig. 4.4b. For the third and fourth instance of finding the optimal hyperplane, it separates class one and class four from the previous steps as shown in Fig. 4.4c, d.

4.2 Affecting Parameters of the SVM Classifier

Two factors that have an influence on SVM classifier performance are training datasets and SVM training parameters. The training dataset should be uniquely represented to be a sample from the focus class. Normally, training data is obtained from a field survey or professional visual interpretation. Aside from the training dataset, SVM training parameters also affect the classifier performance. The following are

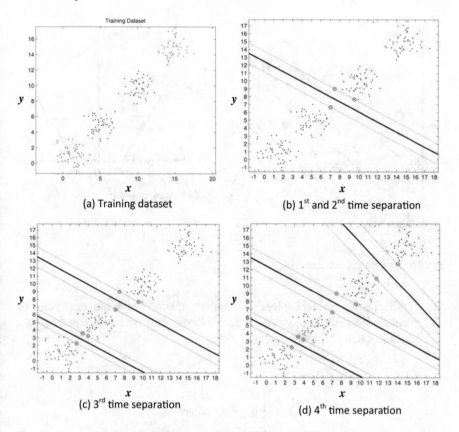

(a) Training dataset

(b) 1st and 2nd time separation

(c) 3rd time separation

(d) 4th time separation

Fig. 4.4 One-against-all technique for separating multiclass classification

three relevant parameters in the SVM training model (Chen 2007; Pai 2006; Wei et al. 2013).

(1)   The kernel function is used to map the non-linear data into linear aspects (Sewell, 2005). There are several kinds of kernel functions; for example, polynomial and hyperbolic tangents. In this research, a Gaussian radial basis function kernel was applied and can be expressed as (4.26):

$$K\left(s_i, s_j\right) = \exp\left(-\frac{\left\|s_i - s_j\right\|^2}{2\sigma^2}\right) \tag{4.26}$$

where $s_i \in \Re^N$ is an N-dimensional data vector for each sample belonging to a class labeled as y_i. The standard deviation (σ) defines the kernel function in this research.

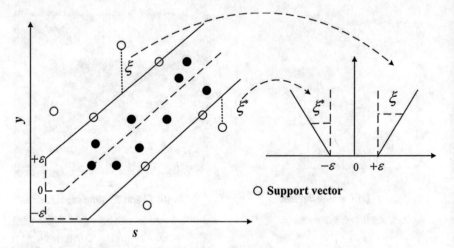

Fig. 4.5 ε-Insensitive loss function

(2) ε, which is the insensitive loss function ($|\xi|_\varepsilon$), is related to the approximation accuracy (Sewell). The loss equals zero if the forecast value is within the ε-tube illustrated in (4.27) and Fig. 4.5.

$$|\xi|_\varepsilon = \begin{cases} 0, & |\xi| \le \varepsilon \\ |\xi| - \varepsilon, & \text{otherwise} \end{cases} \tag{4.27}$$

(3) The regularization parameter (C) defines the trade-off cost between minimizing the training error and the model's complexity (Sewell).

4.3 Summary

SVM is an advanced classification algorithm, and various researchers have applied the SVM algorithm in their works. In this book, the SVM algorithm was explained for constructing an improved algorithm for classification in the next chapter, which has slightly enhanced performance for flood identification in the study area. For this chapter, the SVM algorithm for binary classes and multiclasses was described. The latter half of this chapter discussed the parameters used in an SVM which affect the classification performance.

References

Boswell D (2002) Introduction to support vector machines

Chen KY (2007) Forecasting systems reliability based on support vector regression with genetic algorithms. Reliab Eng Syst Saf 92(4):423–432

Fletcher T (2009) Support vector machines explained. http://www.tristanfletcher.co.uk/SVM%20Explained.pdf

Huang C, Davis LS, Townshend JRG (2002) An assessment of support vector machines for land cover classification. Int J Remote Sens 23(4):725–749

Lindsay D (2003) Matlab workshop 2 an introduction to support vector machine implementations in matlab. Available from: http://www.robots.ox.ac.uk/~az/lectures/ml/matlab2.pdf

Pai P-F (2006) System reliability forecasting by support vector machines with genetic algorithms. Math Comput Model 43(3–4):262–274

Theodoridis S, Koutroumbas K (2008) Pattern recognition, 4th edn. Academic Press

Theodoridis S, Pikrakis A, Koutroumbas K, Cavouras D (2010) Introduction to pattern recognition: a Matlab approach. Academic Press

Venturad (2009) Svm example. http://axon.cs.byu.edu/Dan/678/miscellaneous/SVM.example.pdf

Watanachaturaporn P, Varshney PK, Arora MK (2005) Multisource fusion for land cover classification using support vector machines. 2005 8th International Conference on Information Fusion

Wei Z, Tao T, Ding ZS, Zio E (2013) A dynamic particle filter-support vector regression method for reliability prediction. Reliab Eng Syst Saf 119:109–116

Chapter 5
Improved Support Vector Machine Classifier Through a Particle Filter Algorithm

Recently, state estimation techniques are widely used in many systems. For instance, a tracking system uses the state estimation technique to obtain the exact position of an object. Some other works apply it to estimated parameters before feeding those parameters into their considered system. The aim of this chapter is to understand a constructed state estimation as an advantage in constructing the Support Vector Machine based on a Particle Filter (SVM-PF) classification method. The fundamentals of probability and differential equations are critical concepts in this chapter.

5.1 Linear Dynamics Systems

"Dynamic system" represents all attributes of interest of the system varying with time (Grewal and Andrews 2001; Zarchan and Musoff 2009). Since the seventeenth century, differential equations have become a mathematical model for various dynamic systems. However, it is important to note that it is not all dynamic systems that are modeled by differential equations. One example of a dynamic system is the motion of the planets in the solar system, the movement of vehicles, and the change of the Normalized Different Vegetation Index (NDVI) in an ecological system. In general, the problems which may be solved by any methods should be finite in terms of time. Of special note in the state estimation technique is to call differential equations and their solution, the 'state space approach', with the dependent variable of interest in the differential equation called the state variable of the dynamic system. In addition, another fundamental aspect to form the dynamic system is the time variant and time invariant concept. Time invariant represents the function that does not depend on time, whereas time variant is any function that varies with a change in time.

© Higher Education Press and Springer Nature Singapore Pte Ltd. 2021   57
C. Cao et al., *Environmental Remote Sensing in Flooding Areas*,
https://doi.org/10.1007/978-981-15-8202-8_5

5.1.1 Linear Continuous Systems

The first step is to consider the system in terms of a continuous or discrete system. For continuous functions herein are functions that are not restricted by points in time and are continuous over some real interval $t \in [t_0, t_f]$, while the discrete function can represent a particular value at some point in time with the discrete set of time $t \in \{1, 2, 3, \cdots\}$. Table 5.1 represents mathematical models of continuous linear dynamic systems.

The first-order differential equation is represented by $x(t)$, where the notation 'dot' signifies the derivative with respect to time. The first-order differential equation also may be called a linear differential equation. Many systems can be explained by the general form of a time-varying differential equation as presenting in Table 5.1 which are state equations of the dynamic system. The state variables of the dynamic system are illustrated in Eq. (5.1), in which they are kept into a single n-vector (or state vector).

$$x(t) = [x_1(t)x_2(t)x_3(t) \cdots x_n(t)] \tag{5.1}$$

Moreover, n-dimensional of the state vector is called the state space of the dynamic system. Other parameters' definitions are demonstrated in Table5.2.

All of these can model the dynamic nature of a system expressed by the equation

$$x(t) = F(t)x(t) + C(t)u(t) \tag{5.2}$$

In practice, only the input and output of the system can be measured. $z(t)$ is considered as the measured valued related to the state variables and the inputs by:

$$z(t) = H(t)x(t) + D(t)u(t) \tag{5.3}$$

The definition of each parameter is explained in Table 5.3.

Table 5.1 Mathematical models of continuous linear dynamic system

Time invariant	
Linear	$x(t) = Fx(t) + Cu(t)$
General	$\dot{x}(t) = f(x(t), u(t))$
Time varying	
Linear	$\dot{x}(t) = F(t)x(t) + C(t)u(t)$
General	$x(t) = f(t, x(t), u(t))$

Table 5.2 Explanation of dynamic coefficient matrix, input coupling matrix, and input vector

Name	Component of matrices
Dynamic coefficient matrix (dynamic matrix)	$$F(t) = \begin{bmatrix} f_{11}(t) & f_{12}(t) & f_{13}(t) & \cdots & f_{1n}(t) \\ f_{21}(t) & f_{22}(t) & f_{23}(t) & \cdots & f_{2n}(t) \\ \vdots & \vdots & \vdots & & \vdots \\ f_{n1}(t) & f_{n2}(t) & f_{n3}(t) & \cdots & f_{nn}(t) \end{bmatrix}$$
Input coupling matrix	$$C(t) = \begin{bmatrix} c_{11}(t) & c_{12}(t) & c_{13}(t) & \cdots & c_{1r}(t) \\ c_{21}(t) & c_{22}(t) & c_{23}(t) & \cdots & c_{2r}(t) \\ \vdots & \vdots & \vdots & & \vdots \\ c_{n1}(t) & c_{n2}(t) & c_{n3}(t) & \cdots & c_{nr}(t) \end{bmatrix}$$
Input vector	$u(t) = [u_1(t) u_2(t) u_3(t) \cdots u_r(t)]^{\mathrm{T}}$

Table 5.3 Explanation of measurement sensitivity matrix, output coupling matrix, and measurement vector

Name	Component of matrices
Measurement sensitivity matrix	$$H(t) = \begin{bmatrix} h_{11}(t) & h_{12}(t) & h_{13}(t) & \cdots & h_{1n}(t) \\ h_{21}(t) & h_{22}(t) & h_{23}(t) & \cdots & h_{2n}(t) \\ \vdots & \vdots & \vdots & & \vdots \\ h_{l1}(t) & h_{l2}(t) & h_{l3}(t) & \cdots & h_{ln}(t) \end{bmatrix}$$
Output coupling matrix	$$D(t) = \begin{bmatrix} d_{11}(t) & d_{12}(t) & d_{13}(t) & \cdots & d_{1r}(t) \\ d_{21}(t) & d_{22}(t) & d_{23}(t) & \cdots & d_{2r}(t) \\ \vdots & \vdots & \vdots & & \vdots \\ d_{l1}(t) & d_{l2}(t) & d_{l3}(t) & \cdots & d_{lr}(t) \end{bmatrix}$$
Measurement vector (output vector)	$z(t) = [z_1(t)\, z_2(t)\, z_3(t) \cdots z_l(t)]^{\mathrm{T}}$ $z(t) = [z_1(t) z_2(t) z_3(t) \cdots z_r(t)]^{\mathrm{T}}$

5.1.2 Discrete Linear Systems

The linear differential equation is used to represent the system in the continuous linear system while difference equations are used for the discrete-time version. According to Table 5.4, x_k or $x(t_k)$ stand for the sequence of values of the state variable x, and $x_i(t_k)$ is used for addressing a particular component at a given time. Furthermore, x_k depends upon x_{k-1} and u_{k-1}.

Where

Φ_k: The state transition matrix (STM)

Table 5.4 Mathematical models of discrete linear dynamic system

Time invariant	
Linear	$x_k = \Phi x_{k-1} + \Gamma u_{k-1}$
General	$x_k = f(x_{i-1}, u_{k-1})$
Time-variant	
Linear	$x_k = \Phi_{k-1} x_{k-1} + \Gamma_{k-1} u_{k-1}$
General	$x_k = f(k, x_{k-1}, u_{k-1})$

Γt_k: The input coupling matrix

From these notations, the discrete-time state equations represents the system and measurement models of Eqs. (5.4) and (5.5) respectively.

$$x_k = \Phi_{k-1} x_{k-1} + \Gamma_{k-1} u_{k-1} \tag{5.4}$$

$$z_k = H_k x_k + D_k u_k \tag{5.5}$$

5.2 Random Process and Stochastic Systems to Model State Estimation

Normally uncertainties are present in every system. The uncertainties in the dynamic processes and the measurement are formed based on random processes and stochastic systems. Properties of uncertain dynamic systems are characterized by statistical parameters such as means, correlations, and covariance. These numerical parameters depend on statistical properties such as orthogonality, stationarity, ergodicity, Markovianness of the random processes, Gaussianity of probability distributions, the autocorrelation functions, and power spectral densities of such processes (Grewal and Andrews 2001; Ross 2004). However, this dissertation briefly explains and only addresses the statistical parameters and properties which are important to model the dynamic system. Before dealing with random processes and their properties, a random variable is described.

5.2.1 Probability, Random Variables, and Their Statistical Properties

In the interest of giving a clear explanation, random variables are explained together with an example. In the tossing of a die, one toss gives one "outcome" which corresponding to one of the six faces. Let us label these possible outcomes as π_a, π_b,

π_c, π_d, π_e, π_f. The set of all possible outcomes is called a "sample space" which defined as $\Omega = \{\pi_a, \pi_b, \pi_c, \pi_d, \pi_e, \pi_f\}$. A "random variable" assigns a real number to outcomes. For instance, this defines a 'dot' function $d: \Omega \to \Re$ on the sample space Ω, where $d(\pi)$ is the number of dots representing the result π of the experiment.

$$d(\pi_a) = 1 \quad d(\pi_b) = 2 \quad d(\pi_c) = 3 \quad d(\pi_d) = 4 \quad d(\pi_e) = 5 \quad d(\pi_f) = 6$$

This feature is an example of a random variable.

The expression of $E\langle f(x)\rangle$ determines the expected value of the function f applied to the ensemble of possible values of the random variable x. In addition, a "function of a random variable" x is used to explain the operation of assigning each value of x to another value; for example, the Eq. (5.6)

$$y = f(x) \tag{5.6}$$

where x and y are referred to as input and output respectively.

Moreover, the probability distribution of the average score from tossing n dice tends toward a "Gaussian distribution." A probability distribution with a density function for every real value of x has the notation $N(\bar{x}, \sigma^2)$ and denoted by:

$$p(x) = \frac{1}{\sqrt{2\pi}\sigma} \exp\left[-\frac{1}{2}\frac{(x - \bar{x})^2}{\sigma^2}\right] \tag{5.7}$$

where

$$\bar{x} = E\langle x\rangle \tag{5.8}$$

This is the mean of the distribution. σ^2 is its variance and N stands for "normal", another name of Gaussian distribution. Another fundamental theory of random variables is joint probability and conditional probabilities. The relationship between two random variables can frequently be clarified by considering the conditional distribution of one event given the value of another event. For any two events χ_a and χ_b, the conditional probability of the χ_a given χ_b is defined by:

$$p(\chi_a|\chi_b) = \frac{p(\chi_a \cap \chi_b)}{p(\chi_b)} \tag{5.9}$$

$p(\chi_a \cap \chi_b)$ is the joint probability of χ_a and χ_b which the joint probability of independent events is the product of their probabilities when $p(\chi_b) > 0$. The Eq. (5.9) is known as Bayes' rule.

5.2.2   Statistical Properties of Random Processes and Random Sequence

A function $x(s)$ defines a random variable for each outcome of an experiment identified as S. If we assign to each outcome S a time function $x(t, s)$, then we obtain a series
of functions called "random processes" or "stochastic processes." A random process
is discrete if its argument is a discrete variable, and is called a random sequence.

$$x(k, s)   k = 1, 2, 3 \cdots \tag{5.10}$$

The value of a random process $x(t)$ at any particular time $t = t_0$, namely $x(t_0, s)$,
is a random variable. Letting $x(t)$ be an n-dimensional vector random process, the
mean value is defined by:

$$Ex(t) = \int_{-\infty}^{\infty} x(t) p[x(t)] dx(t) \tag{5.11}$$

Alternatively, for a random sequence the expected value is represented by:

$$Ex_i(t) = \sum_{i=1}^{n} x_i(t) p[x_i(t)] dx(t), \quad i = 1, 2, \cdots, n \tag{5.12}$$

The correlation of $x(t)$ is defined by:

$$E\langle x(t_1) x^T(t_2)\rangle = \begin{bmatrix} E\langle x_1(t_1) x_1(t_2)\rangle & \cdots & E\langle x_1(t_1) x_n(t_2)\rangle \\ \vdots & & \vdots \\ E\langle x_n(t_1) x_1(t_2)\rangle & \cdots & E\langle x_n(t_1) x_n(t_2)\rangle \end{bmatrix} \tag{5.13}$$

where

$$Ex_i(t_1) x_j(t_2) = \int_{-\infty}^{\infty} \int x_i(t_1) x_j(t_2) p[x_i(t_1), x_j(t_2)] dx_i(t_1) x_j(t_2) \tag{5.14}$$

The covariance of $x(t)$ is defined by

$$E\langle [x(t_1) - Ex(t_1)][x(t_2) - Ex(t_2)]^T\rangle = E\langle x(t_1) x^T(t_2)\rangle - E\langle x(t_1)\rangle E\langle x^T(t_2)\rangle \tag{5.15}$$

If the process $x(t)$ has a mean of zero, its correlation and covariance are equal.
Moreover, the random process $x(t)$ is called uncorrelated if:

$$E\langle [x(t_1) - E\langle x(t_1)\rangle][x(t_2) - E\langle x(t_2)\rangle]^T\rangle = Q(t_1, t_2)\delta(t_1 - t_2) \tag{5.16}$$

where $\delta(t)$ is the Direc delta function, expressed as:

$$\int_a^b \delta(t)dt = \begin{cases} 1 & if \ a \le 0 \le b, \\ 0 & otherwise \end{cases} \tag{5.17}$$

In the case of a random sequence, x_k is called uncorrelated if:

$$E\left\langle [x_k - E\langle x_k\rangle][x_j - E\langle x_j\rangle]^T \right\rangle = Q(k, j)\Delta(k - j) \tag{5.18}$$

where $\Delta(\cdot)$ is the Kronecker delta function, expressed as:

$$\Delta(k) = \begin{cases} 1 & if \ k = 0, \\ 0 & otherwise \end{cases} \tag{5.19}$$

A white-noise process or sequence is an example of an uncorrelated process or sequence. In the case of two random processes, the correlation matrix of two random processes $x(t)$, an n-vector, and $y(t)$, an m-vector, is given by a $n \times m$ matrix.

$$Ex(t_1)y^T(t_2) \tag{5.20}$$

Similarly, the cross-covariance $n \times m$ matrix is

$$E\left\langle [x(t_1) - Ex(t_1)][y(t_2) - Ey(t_2)]^T \right\rangle \tag{5.21}$$

For two random processes $x(t)$ and $y(t)$, they are called uncorrelated if their cross-covariance matrix is equal to zero for all t_1 and t_2:

$$E\left\langle [x(t_1) - Ex(t_1)][y(t_2) - Ey(t_2)]^T \right\rangle = 0 \tag{5.22}$$

For the two random processes $x(t)$ and $y(t)$ are called orthogonal, if their correlation matrix is equal to zero for all t_1 and t_2:

$$x(t_1)y^T(t_2) = 0 \tag{5.23}$$

Other statistical properties of random processes are further described by (Ross 2004).

5.2.3 Linear System Models of Random Processes and Random Sequences

Let a linear system be given by

$$y(t) = \int_{-\infty}^{\infty} x(\tau)h(t, \tau)d\tau \tag{5.24}$$

Fig. 5.1 Block diagram representation of a linear system

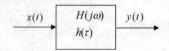

$x(t)$ is the input and $h(t, \tau)$ is the system weighting function. If the system is time invariant, then the linear system becomes

$$y(t) = \int_{-\infty}^{\infty} h(\tau)x(t - \tau)\mathrm{d}\tau \qquad (5.25)$$

This type of integral is called a convolution integral.

According to Fig. 5.1, $h(\tau)$ is a transfer function in time-domain, and $H(j\omega)$ is a transfer function in frequency-domain, which has relationships based on a Laplace transformation such as

$$H(s) = \int_{0}^{\infty} h(\tau)e^{s\tau}\mathrm{d}\tau \qquad (5.26)$$

where $s = j\omega$

5.2.3.1   Stochastic Differential Equations for Random Processes

A linear stochastic differential equation as a form of random process has an initial condition which is

$$x(t) = F(t)x(t) + G(t)w(t) + C(t)u(t)$$
$$z(t) = H(t)x(t) + v(t) + D(t)u(t) \qquad (5.27)$$

where

$x(t) = n \times 1$ state vector
$z(t) = l \times 1$ measurement vector
$u(t) = r \times 1$ deterministic input vector
$F(t) = n \times n$ time-varying dynamic coefficient matrix
$C(t) = n \times r$ time-varying input coupling matrix
$H(t) = l \times n$ time-varying measurement sensitivity matrix
$D(t) = l \times r$ time-varying output coupling matrix
$G(t) = n \times r$ time-varying process noise coupling matrix
$w(t) = r \times 1$ zeromean uncorrelated "system noise" process
$v(t) = l \times 1$ zeromean uncorrelated "measurement noise" process

Moreover, the expected values are

$$E\langle w(t)\rangle = 0$$

$$E\langle v(t)\rangle = 0$$
$$E\langle w(t_1)w^T(t_2)\rangle = Q(t_1)\delta(t_2 - t_1)$$
$$E\langle v(t_1)v^T(t_2)\rangle = R(t_1)\delta(t_2 - t_1)$$
$$E\langle w(t_1)v^T(t_2)\rangle = M(t_1)\delta(t_2 - t_1)$$

The dimensions of matrices Q, R, M are $r \times r$, $l \times l$, $r \times l$ respectively. The function $u(t)$ normally represents a known control input. For this dissertation, it will be assumed that $u(t) = 0$.

5.2.3.2   Stochastic Difference Equations for Random Sequence

For a discrete model of a random sequence the initial conditions can be given by:

$$x_k = \Phi_{k-1}x_{k-1} + G_{k-1}w_{k-1} + \Gamma_{k-1}u_{k-1}$$
$$z_k = H_k x_k + v_k + D_k u_k \tag{5.28}$$

where

$x_k = n \times 1$ state vector
$z_k = l \times 1$ measurement vector
$u_k = r \times 1$ the deterministic input vector
$w_{k-1} = r \times 1$ zero mean uncorrelated "system noise" process
$\Phi_{k-1} = n \times n$ time-varying matrix
$G_{k-1} = n \times r$ time-varying matrix
$H_k = l \times n$ time-varying matrix
$D_k = l \times r$ time-varying matrix
$\Gamma_{k-1} = n \times r$ time-varying matrix

In addition, the expected values are

$$E\langle w_k\rangle = 0,$$
$$E\langle v_k\rangle = 0,$$
$$E\langle w_{k_1} w_{k_2}^T\rangle = Q_{k_1}\Delta(k_2 - k_1)$$
$$E\langle v_{k_1} v_{k_2}^T\rangle = R_{k_1}\Delta(k_2 - k_1)$$
$$E\langle w_{k_1} v_{k_2}^T\rangle = M_{k_1}\Delta(k_2 - k_1)$$

5.3 Particle Filter Algorithm

Even though the Kalman filter (KF) had been used to solve estimation problems that typically are linear systems, for nonlinear systems the KF has difficulty solving them. In such a system, the extended Kalman filter (EKF) has been introduced and is widely used in many applications. However, when the nonlinearities are high, the EKF may restrict with Gaussian assumption, because it implements linearization to propagate the mean and covariance of the state estimate. To solve these problems and to apply it to the system, which does not understand its model, the particle filter (PF) is introduced, which it is a probability-based estimator. For mathematical derivation of the PF, suppose the nonlinear system and measurement equations at time point k are given as:

$$x_{k+1} = f_k(x_k) + w_k \tag{5.29}$$

$$z_k = h_k(x_k) + v_k \tag{5.30}$$

$x_k \in \mathfrak{R}^{n_x}$ is the state, w_k is the noise in the process, z_k is the measurement, and v_k is the measurement noise.

Theorem 1 An alternative formulation of the state-space model is as follows

$$x_{k+1} \sim p(x_{k+1}|x_k)$$
$$z_k \sim p(z_k|x_k)$$

If the dynamic model is given by the above equations, the filtering density $p(x_k|z_{1:k})$ and the one-step-ahead prediction density $p(x_{k+1}|z_{1:k})$ are given by:

$$p(x_k|z_{1:k}) = \frac{p(z_k|x_k)p(x_k|z_{1:k-1})}{p(z_k|z_{1:k-1})}$$

$$p(x_k|z_{1:k}) = \int_{\mathfrak{R}^{n_x}} p(x_{k+1}|x_k)p(x_k|z_{1:k})dx$$

where

$$p(z_k|z_{1:k-1}) = \int_{\mathfrak{R}^{n_x}} p(z_k|x_k)p(x_k|z_{1:k-1})dx$$

Based on Eqs. (5.29) and (5.30) using Bayes theorem and Markov property the "filtering density" can be written as:

$$p(x_k|z_{1:k}) = \frac{p(z_k|x_k)p(x_k|z_{1:k-1})}{p(z_k|z_{1:k-1})} \tag{5.31}$$

Thus,

$$p(x_k|z_{1:k}) \propto \underbrace{p(z_k|x_k)}_{\omega(x_k)} \underbrace{p(x_k|z_{1:k-1})}_{q(x_k)} \tag{5.32}$$

The aim of the PF(Arulampalam et al. 2002; Gordon et al. 1993; Jia 2015; Johannes and Polson 2009; Salmond and Gordon 2005; Zhang 2011) is to find a recursive way to compute the conditional probability density function $p(x_k|z_{1:k})$, where $q(x_k)$ is the importance density and $\omega(x_k)$ is the importance weight. At the initial state ($k = 0$), starting with initializing the particles and their relative weights:

$$x_0^i \sim p(x_0), \quad i = 1, \cdots, N \tag{5.33}$$

$$\omega_0^i \sim \frac{1}{N}, \quad i = 1, \cdots, N \tag{5.34}$$

N is the number of particles, which results in the following approximation:

$$\widehat{P}_N(x_0) = \sum_{i=1}^{N} \frac{1}{N} \delta(x_0 - x_0^i) \tag{5.35}$$

$\delta(\cdot)$ is the Direc delta, which has zero value everywhere, except at $x_0 - x_0^i$, where it is infinitely large such that its total integral is 1. At time k, assume the following approximation:

$$\widehat{P}_N(x_{k-1}|z_{1:k-1}) = \sum_{i=1}^{N} \omega_{k-1}^i \delta(x_{k-1} - x_{k-1}^i) \tag{5.36}$$

This is available from time $k - 1$. Now, the task is to approximate $p(x_k|z_k)$ using the importance sampling principle in which the importance density is given by $q(x_k|x_{k-1}, z_k)$. Therefore, N samples can be generated according to:

$$\tilde{x}_k^i \sim q(x_k|x_{k-1}^i, z_k), \quad i = 1, \cdots, N \tag{5.37}$$

In addition, the weights are computed according to:

$$\tilde{\omega}_k^i = \frac{p(\tilde{x}_k^i|z_{1:k-1})}{q(\tilde{x}_k^i|x_{k-1}^i, z_k)}, \quad i = 1, \cdots, N \tag{5.38}$$

According to the theorem of the dynamic model (Theorem 1), it can be stated that:

$$p(x_k|z_{1:k-1}) = \int_{\Re^{n_x}} p(x_k|x_{k-1})p(x_{k-1}|z_{1:k-1})dx \tag{5.39}$$

Applying the assumption in Eqs. (5.36), (5.38), and (5.39) it generates

$$\widetilde{\omega}_k^i = \frac{\int_{\mathfrak{R}^{n_x}} p\big(\tilde{x}_k^i | x_{k-1}\big) \sum_{i=1}^{N} \omega_{k-1}^i \delta\big(x_{k-1} - x_{k-1}^i\big) dx_{k-1}}{q\big(\tilde{x}_k^i | x_{k-1}^i, z_k\big)} \tag{5.40}$$

$$\widetilde{\omega}_k^i = \frac{p\big(\tilde{x}_k^i | x_{k-1}^i\big)}{q\big(\tilde{x}_k^i | x_{k-1}^i, z_k\big)} \omega_{k-1}^i \tag{5.41}$$

Thus, the approximation of the one-step-ahead prediction can be obtained:

$$\widehat{P}_N(x_k | z_{1:k-1}) = \sum_{i=1}^{N} \widetilde{\omega}_k^i \delta\big(x_k - \tilde{x}_k^i\big) \tag{5.42}$$

where the particles are generated according to the Eq. (5.37) and their weights given by the Eq. (5.41). To obtain an approximation of $p(x_k | z_{1:k})$ in Eq. (5.31), we need to find $p(z_k | z_{1:k-1})$, and according to Theorem 1:

$$p(z_k | z_{1:k}) = \int p(z_k | x_k) p(x_k | z_{1:k-1}) dx_k \tag{5.43}$$

From Eqs. (5.42) and (5.43), it is implied that:

$$\widehat{P}_N(z_k | z_{1:k-1}) = \int p(z_k | x_k) \sum_{i=1}^{N} \widetilde{\omega}_k^i \delta\big(x_k - \tilde{x}_k^i\big) dx$$
$$\widehat{P}_N(z_k | z_{1:k-1}) = \sum_{i=1}^{N} p\big(z_k | \tilde{x}_k^i\big) \widetilde{\omega}_k^i \tag{5.44}$$

Therefore, inserting Eqs. (5.42) and (5.44) into (5.31) produces:

$$\widehat{P}_N(x_k | z_{1:k}) = \sum_{i=1}^{N} \frac{p\big(z_k | \tilde{x}_k^i\big) \widetilde{\omega}_k^i}{\sum_{j=1}^{N} p\big(z_k | \tilde{x}_k^j\big) \widetilde{\omega}_k^j} \delta\big(x_k - \tilde{x}_k^i\big) dx \tag{5.45}$$

Finally, the conditional probability density function can be obtained, and the procedure is concluded in the Table 5.5.

Figure 5.2 Presents workflow propagation of the particles of the PF algorithm in an estimated state vector at any epoch. One important step that has an influence on the PF processes is the re-sampling step, which will be described in the following section.

The re-sampling step is important for the PF to work, because without re-sampling the variance of importance weights will grow to infinity and the filter will diverge. The re-sampling step removes low importance weight particles and produces multiple copies of particles with high importance weights (Ho et al. 2006; Schön 2010; Turner 2013). The re-sampling step has a role in changing the non-uniformly weighted density into a uniformly weighted density. The approximated distribution produced by the PF before re-sampling in given by:

$$\widehat{P}_N(x_k | z_{1:k}) = \sum_{i=1}^{N} \omega_k^i \delta\big(x_k - \tilde{x}_k^i\big) \tag{5.46}$$

Table 5.5 Summarized particle filter

(1) Initialize the particles $x_0^i \sim p(x_0)$, $i = 1, \cdots, N$ and the weights, $\omega_0^i \sim \frac{1}{N}$, $i = 1, \cdots, N$ and Let $k = 1$.

(2) Generate N new particles by drawing from the importance density

$$\tilde{x}_k^i \sim q(x_k | x_{k-1}^i, z_k), \quad i = 1, \cdots, N$$

Update the weights accordingly

$$\tilde{\omega}_k^i = \frac{p(\tilde{x}_k^i | x_{k-1}^i)}{q(\tilde{x}_k^i | x_{k-1}^i, z_k)} \omega_{k-1}^i, \quad i = 1, \cdots, N$$

(3) Further update the weights according to,

$$\tilde{\omega}_k^i = \frac{p(z_k | x_k^i) \tilde{\omega}_k^i}{\sum_{j=1}^{N} p\left(z_k | x_k^j\right) \tilde{\omega}_k^j}, \quad i = 1, \cdots, N$$

(4) Resample according to the re-sampling algorithm in the following section

(5) Set $k = k + 1$ and Iterate from step (2).

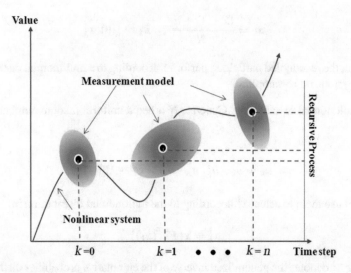

Fig. 5.2 Particle filter algorithm workflow

After re-sampling the approximated distribution is given by:

$$\hat{P}_N(x_k | z_{1:k}) = \sum_{i=1}^{N} \frac{1}{N} \delta\left(x_k - x_k^i\right) \tag{5.47}$$

This is accomplished by drawing a new sample x_k^i as a replacement for each $i = 1, \cdots, N$ according to

$$P\left(x_k^i = \tilde{x}_k^j\right) = \omega_k^j, \quad j = 1, \cdots, N \tag{5.48}$$

These re-sampling algorithms can be summarized as:

(1) Systematic re-sampling: Generate N ordered numbers

$$u_k = \frac{(k-1)+\tilde{u}}{N}, \tilde{u} \sim U[0, 1]$$

The re-sampled particles are obtained by producing n_i copies of particle x_k^i, where

$$n_i = \text{the number of } u_k \in \left(\sum_{s=1}^{i-1} \omega_k^s, \sum_{s=1}^{i} \omega_k^s \right]$$

(2) Stratified re-sampling: Generate N ordered numbers

$$u_k = \frac{(k-1)+\tilde{u}_k}{N}, \quad \tilde{u}_k \sim U[0, 1]$$

Then, the re-sampled particle x_k^i performs according to a multinomial distribution as the systematic re-sampling.

(3) Multinomial re-sampling: Generate N ordered uniform random numbers

$$u_k = u_{k+1}\tilde{u}_k^{\frac{1}{k}}, u_N = \tilde{u}_N^{\frac{1}{N}} \ \tilde{u}_k \sim U[0, 1]$$

Then use them to select x_k^i according to the multinomial distribution in:

$$x_k^i = x\left(F^{-1}(u_k)\right)$$

where F^{-1} denotes the generalized inverse of the cumulative probability distribution of the normalized particle weights.

(4) Residual re-sampling: allocate $n_i' = [N\omega_i]$ copies of the particle x_i to the new distribution. Additionally, resample $m = N - \sum_{i=1}^{N} n_i'$ particles from $\{x_i\}$ by making n_i'' copies of particle x_i where the probability for selecting x_i is proportional to $\omega_i' = N\omega_i - n_i'$ using one of the mentioned re-sampling schemes earlier.

The implementation method in this dissertation, the systematic re-sampling, is employed in the proposed method (a support vector machine-based particle filter method), which is used in the enhancement of water identification in a flooding area and in improved class separation in the wetland area.

5.4 A Support Vector Machine-Based Particle Filter (SVM-PF)

The state estimation technique can improve classification performance by searching some parameters that may be more suitable to the system or dataset in which it can enhance the correlation between dataset and SVM training parameters. A strong correlation results in an extremely favorable performance, whereas a weak correlation results in an inadequate SVM model. Parameter estimation is needed to improve the correlation. Parameter value selection by PF can avoid a biased setting and produce rationally selected parameter values based on the particle weights that result from the Probability Density Function (PDF) calculation. Frequently, a PF is implemented in a dynamic system to estimate the states. A PF consists of two important components: using prediction and updating synergistically and functions as an iterative technique. Using these two components a state vector composed of the estimated parameters is enabled to update its value with each iteration based on the weight of each particle calculated by the PDF of the output of the prediction process. Thus, a true measurement value from an observation system is obtained.

To implement a PF for an estimated parameter, we can create the true measurement sequence for the updating process by imitating an observation value that is equal to 1 (100%). The final updated state vector will be the input of the training process of the SVM. The SVM-PF-improved flood classification approach beginning with a PF is implemented to estimate the SVM training parameters by retaining the SVM training function as a measurement mode. It uses a normal distribution that describes the fluctuation of the SVM parameter values as defined in the system model. Each iteration of the PF process provides updated parameters to obtain reasonable, unbiased values. Finally, the SVM-PF classification model results from the SVM training process input into the updated parameters. To estimate the parameters of the SVM training model by PF, the system and measurement model are defined as follows:

$$x_{k+1} = x_k + w_k \tag{5.49}$$

$$y_k = h_k(x_{k|k-1}) + v_k \tag{5.50}$$

where $x_k = [\sigma_k, C_k, \varepsilon_k]$ represents the state vector at a time k. w_k isthe nonlinear noise with a mean of zero and variance Q, which explains the fluctuations of the SVM parameters.y_k is the true measurement vector, or the accuracy associated with the SVM training function h_k, and v_k is the nonlinear prediction error with a mean of zero and variance R. A flowchart of the SVM-PF method is depicted in Fig. 5.3, and the main steps are as follows:

1. The initial values of the SVM parameter vector elements are set as x_0. "n" particles are set, and their weights are established as $\{x_0^i = x_0, w_0^i = 1/n\}_{i=1}^n$.
2. For $k = 1,2,3\ldots$

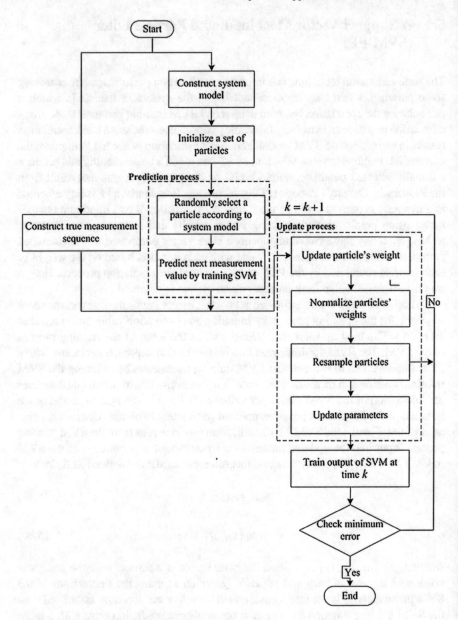

Fig. 5.3 Flowchart of the proposed (SVM-PF) method

(1) Use the particle set $\left\{\hat{x}_{k-1}^{j}, 1/n\right\}_{i=1}^{n}$ from the previous time $k-1$ with the equation in system model (5.49).

(2) Predict the SVM output measurement $\hat{y}_{k|k-1}^{j}$ by training the SVM with the state vector $\hat{x}_{k-1}^{j}$ using (5.50).

(3) Create the true measurement sequence y_k by duplicating the observation value of the SVM training model performance $y_k = 1$.

(4) Using the true measurement y_k value update each particle's weight as $\omega_k^i = p(y_k|\hat{x}_{k|k-1}^i)$.

(5) Normalize the particle weights as $\omega_k^i = \omega_k^i / \sum_{j=1}^{n} \omega_k^j$.

(6) The particles will be rejected or retained depending on their weight (ω_k^i) after being processed by a re-sampling algorithm.

(7) Obtain the output parameter estimate at time k:

$$\hat{x}_k = \sum_{i=1}^{n} \frac{\hat{x}_k^i}{n} \tag{5.51}$$

3. Train the SVM model with the updated parameter vector $\hat{x}_k$ to find the prediction of $\hat{y}_{k+1}$.

4. Confirm whether to stop iterations based on the pre-defined threshold given by the minimum error values.

The updated parameters or state vector at the last iteration will be used as an input to the SVM training model, which results in the SVM-PF classification model. The stopping criteria of the PF can be defined using various techniques, such as a minimum error setting, and are implemented in this research. Moreover, the correlation in the study at any iterative time between the SVM training model parameters and the dataset is represented by a term representing the average value of every particle's weight at that time, which is shown as an Eq. (5.52).

$$corr_k(parameters, dataset) = \frac{1}{n^2 - n} \sum_{i=1}^{n} \omega_k^i \tag{5.52}$$

5.5 A SVM-PF Applied in the Study Area

The accuracy of the conventional method was employed to be a threshold for the SVM-PF method. According to Fig. 5.4, the processes started with the SAR pre-image processing and the sample was done using regions of interest (ROI) such as water and non-water areas designated by professional visual interpretation. The sample was separated into the training and testing dataset equally. The SVM-PF

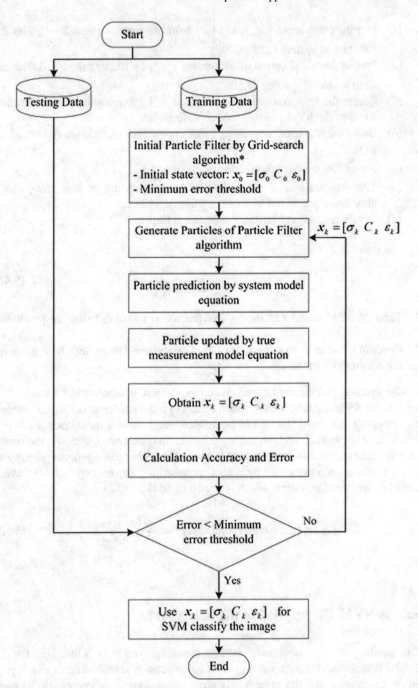

Fig. 5.4 Research workflow of improving flood classification by Particle Filter.
*Initial Particle Filter by Grid-search algorithm presented in Fig. 5.3

Fig. 5.5 Flowchart of the
initial Particle Filter by
Grid-search algorithm

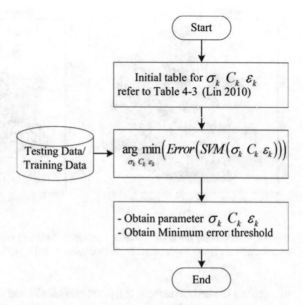

method worked as an iterative technique to select the initial SVM training parameter. The PF would generate those parameter value based on the weight of each particle, which can avoid the impact of human bias. Iterations of the PF to obtain suitable initial values continued until the accuracy of SVM classification model met the predefined threshold. The flowchart of the initial Particle Filter by Grid-search algorithm is shown in Fig. 5.5.

5.6   Measured Results of the SVM-PF in Water Identification of Flooding Area

5.6.1   Reference Dataset

The SAR imagery covers Ayutthaya province as shown in Fig. 5.6a. Pre-image processing is provided by Geo-Informatics and Space Technology Development Agency (GISTDA), Thailand. A sampling dataset for the SVM training model and the SVM-PF training model were selected using visual interpretation to generate a sample size of 888,960 pixels (i.e., 95,104 flood pixels and 793,856 non-flood pixels) based on the flood map reference provided by GISTDA (Fig. 5.6b).

The sampling data were equally divided into training and testing datasets. Each of the datasets had 47,552 flood pixels and 396,928 non-flood pixels. Three types of texture analyses were implemented in the study: mean, entropy, and fuzzy entropy (Khushaba et al. 2011). The experiment implemented block processing of the samples

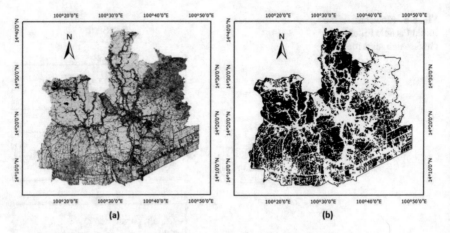

Fig. 5.6 Study areas covering Ayutthaya province, Thailand (**a**) SAR imagery 50-m resolution with ScanSAR narrow mode acquired on December 4, 2011. (**b**) Reference data

to calculate the texture using 8×8 pixel grid blocks and stacked them for subsequent feeding into the classifier.

5.6.2 Accuracy Assessment

In the simulation, the initialized parameter values for the SVM and SVM-PF approaches generated by a Grid-search approach (Hsu et al. 2008) implemented with 10-fold cross-validation produced values of $\sigma = 3.5$, $C = 512$, and $\varepsilon = 0.3$. As mentioned in Chap. 4, the concept of the SVM-PF method, the system model, and true measurement value-building were performed using particles moving in terms of the normal distribution and by imitating the observation value respectively. The number of PF recursions depends on the minimum error values. These values are set to be smaller than the error that occurs in the conventional SVM method in an attempt to demonstrate that the SVM-PF can improve the performance even though the initial parameter values are generated via Grid-search based on k-fold cross-validation, which can normally provide useful results in certain other applications. The measurement results for ten test cases using the SVM and SVM-PF approaches are shown in Table 5.7(Insom et al. 2015).The evaluation is based on accuracy (A), precision (P), and recall (R) (Yuan and Sarma 2011) which are expressed as (5.53), (5.54), and (5.55).

$$A = \frac{tp + tn}{p + n} \tag{5.53}$$

$$P = \frac{tp}{tp + fp} \tag{5.54}$$

Table 5.7 Accuracy, precision, and recall of the proposed method compared with conventional SVM

Test case	Accuracy (%)		Precision (%)		Recall (%)	
	SVM	SVM-PF	SVM	SVM-PF	SVM	SVM-PF
1	99.50	99.78	99.60	99.90	99.84	99.85
2	99.70	99.76	99.79	99.84	99.87	99.89
3	99.57	99.71	99.57	99.69	99.95	99.98
4	99.01	99.71	99.06	99.71	99.84	99.97
5	99.50	99.83	99.50	99.86	99.94	99.95
6	99.68	99.71	99.77	99.79	99.87	99.89
7	99.70	99.71	99.84	99.84	99.82	99.84
8	99.08	99.29	99.03	99.25	99.95	99.97
9	99.58	99.70	99.61	99.73	99.92	99.94
10	99.67	99.70	99.87	99.87	99.76	99.79
Mean	99.50	99.69	99.56	99.75	99.88	99.91
NRMSE	0.056	0.034	0.052	0.031	0.014	0.011

$$R = \frac{tp}{tp + fn} \tag{5.55}$$

Positive sample (p) define the sample of water areas, whereas negative sample (n) represent non-water areas, and t, f represent true and false classified samples respectively. The results show that the SVM-PF method improved the performance in every measurement aspect. In all of the test cases in Table 5.7, the accuracy, precision, and recall of the measurements are higher than those of the SVM method.

The average accuracy, precision, and recall exhibit the same trend: the SVM-PF method reflects an improvement over the SVM method. The normalized root means square errors (NRMSEs) of the ten test cases for these three evaluation aspects are also smaller than those of the original method. Figure 5.7a shows the classification map produced by the conventional SVM method, and Fig. 5.7b illustrates the results generated by the SVM-PF approach. The photographic images were generated from the classification model from test case 4 in Table 5.7, in which highly different classification model performances and highly different z-score measurements for the two methods can be noted (Fig. 5.8a–c).

It can be observed that the classification result generated by the SVM approach displays discrepancies, whereas the water areas are clearly depicted in the image produced by classification using the SVM-PF method. Figure 5.8a–c show the standard score (i.e., the z-score) of the two methods with respect to accuracy, precision, and recall. The SVM-PF produces a z-score that is superior to that of the SVM approach in every assessment and test case. Also, Fig. 5.8d illustrates the variation in correlation as presented in the Eq. (5.52) of another experiment during correlation testing of the SVM-PF approach at every iteration of the PF. The correlation tends to

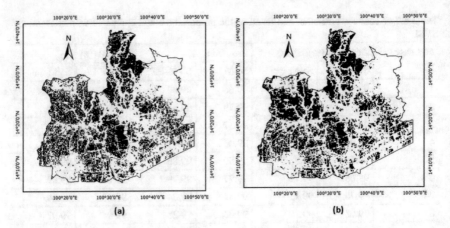

Fig. 5.7 Classified maps result by (**a**) SVM model and (**b**) SVM-PF model from test case 4 in Table 5.7

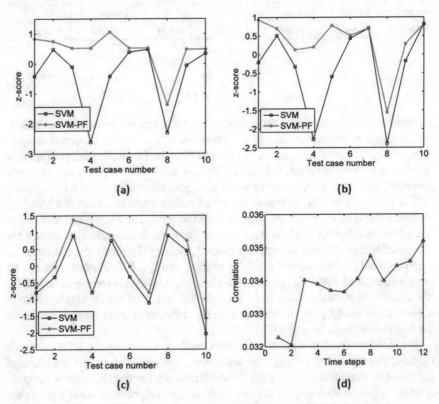

Fig. 5.8 Comparison of z-scores for (**a**) accuracy, (**b**) precision, and (**c**) recall of the SVM and SVM-PF methods. (**d**) The correlation at each iteration of the particle filter

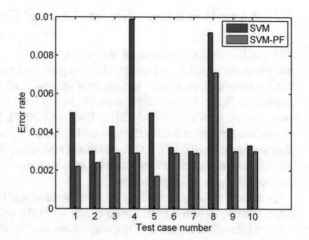

Fig. 5.9 Comparison of error rates of SVM and SVM-PF schemes

improve with further iterations and reaches the highest value at the final time step, thus demonstrating that the PF can achieve appropriate parameter values.

Moreover, the error rates (Yuan and Sarma 2011) presented as Eq. (5.56) for all of the test cases are included in the accuracy assessment (Fig. 5.9).

$$E = \frac{fp + fn}{p + n} \qquad (5.56)$$

In every test case, the SVM method has a higher error rate compared to the SVM-PF approach. An important computational complexity of a Grid-search approach based on k-fold cross-validation and a PF for determining the training parameter values is the time required to calculate the SVM training function. Table 5.8 lists the number of calculations necessary for the training function for both methods. In the table, k is the k value in k-fold cross-validation; M is the number of the setting vectors, which consist of the three parameters; n is the number of particles; and K is the number of iterations of the PF processes. The SVM-PF is more time consuming than the SVM because it requires a substantial number of particles and a recursive process to obtain a result. However, since the advancement of technology in computer development, the extra consumption of time is not such a big issue.

Table 5.8 Comparison of computational complexity of the proposed method compared with the conventional SVM

Approach	SVM training function
SVM	$k \times M$
SVM-PF	$(n \times K) + K$

5.7   An In-House Classifier for All Land Cover—CANFET

The continuous developments of storage capacity and computer processing speed enable advanced machine-learning (ML), in particular supervised ML, which can be used to process large remote sensing data such as satellite imagery. The process of supervised ML is that searching through data to learn patterns can apply what has been learned to new data (Dey 2016; Kotsiantis 2007). In general, supervised ML algorithms have been widely documented in the past literature to be more effective than traditional methods in classification and regression. Nowadays, supervised MLs are different in construction and complexity, which were widely developed depending on specific issues and types of data.

In remote sensing, especially in satellite imaging, the effects of unwanted or impure pixels (referred to as noise) occurring in the data vary with their distribution, and the effect is greater if impurity pixels are included in a classifier's training set. Noise often increases the complexity of classification tasks (Hsieh et al. 2001). A reasonable accuracy may be obtained by classifying a low-noise image with the traditional supervised classification methods (e.g. minimum distance, nearest neighbor, and maximum likelihood). In contrast, as noise increases accuracy may decrease beyond acceptable values. In general, various supervised MLs such as neural networks, support vector machines, and random forests outperform traditional approaches in terms of performance and robustness. However, the main disadvantage of the supervised MLs is that most of them are quite difficult to interpret and sometimes require many parameters.

A new classifier called CANFET (Classification of ANdrews' curves of satellite data using Fuzzy logic and Evidence Theory) (Boonprong et al. 2017) was constructed to overcome the above problems based on Andrews' plots, fuzzy logic, and the Dempster-Shafer theory of evidence (D-S theory). The CANFET surpasses the traditional methods of classification accuracy and has low complexity and low parameters. Its features include pattern visualizing, curve similarity, noise detection, and classification. The explanation of CANFET was divided into three directions. The first is to explain the construction systematically. The core classifier, which is the combination of fuzzification and DS theory, was proposed as well as both of the versions including *Simplified CANFET* and *Direct-matching CANFET*. The second is the example of using CANFETs in many applications. Finally, the proposed algorithm was compared to the performance of both the traditional classification methods and the fundamental machine learning classifiers.

Specifically, Andrews' plots of satellite data can visualize unique patterns for each land cover category due to structural conservation capabilities. Moreover, the mixed pixel and noises can be easily detected by examining the set of Andrews' plots. The finding leads to the inventing of the classifier. The concept of CANFET can be briefly described in the following sentence. We trained each Gaussian membership function by feeding the training curves for each class, and then applied the trained functions to all input curves resulting in a fuzzy membership grade. The final class decision was made based on the maximum *belief* value calculated from the membership grades and Andrews' curve pattern called TAD (Type of Andrews' curve Dynamic) by using the Dempster-Shafer theory of evidence. In the case of the simplified CANFET, it omitted TAD and DS theory of which decisions based on confidence values are superseded by decisions based on the maximum membership value. In the case of the Direct-matching CANFET, the core fuzzification is different from the original CANFET. In contrast to the original CANFET, which calculates membership values from membership functions of a "set of training samples," the direct match CANFET calculates the member value for "every training sample." Additionally, we developed a feature called "confidence zone and weight". It is determining the non-overlapping areas in the training data diagram. This technique is similar to human perception of curve separation because it focuses on the most discriminable area and gives it more weight in the classification.

5.7.1 Important Theories and Concepts

5.7.1.1 Andrews' Plots

Andrews' curves (Andrews 1972) were developed to visualize multi-dimensional data by mapping each observation to a function. For any n bands regardless of their order in a satellite image, an image pixel b can be expressed by $b = \{b_1, b_2, b_3, \cdots, b_n\}$, where t is the time domain from $-\pi$ to π, and the Andrews' function is defined as Eq. (5.57).

$$f_b(t) = b_1/\sqrt{2} + b_2 \sin(t) + b_3 \cos(t) + b_4 \sin(2t) + b_5 \cos(2t) + \cdots \quad (5.57)$$

Each observation is projected onto a set of orthogonal basis functions represented by sines and cosines, and each data point may be viewed as a curve. Because of the mathematical properties of the trigonometric functions, the Andrew's functions preserve the means, distances, and variances. On the one hand, a consequence of this equation is that any samples which have data values in every dimension that are close together produce Andrew's curves that are also close together. On the other hand, if there is a pattern within the data, it may be visible in the Andrews' curve of the data (Cesar and Fyfe 2005). Note that to compare any curves, they must be produced from the same data order.

5.7.1.2 Fuzzification

According to fuzzy set theory, fuzzification is the process that takes the inputs and determines the degree to which they belong to each of the appropriate fuzzy sets via the membership function.

A fuzzy set F based on a universe of discourse (UOD) U with elements u is expressed as Eq. (5.58).

$$F = \int \{\mu_A(u)/u\} \; \forall u \in U \tag{5.58}$$

where $\mu_A(u)$ is a membership function, or FMF, of fuzzy variable u in the set F, and it provides a mapping of U in the closed interval [0,1]. The term $\mu_A(u)$ is simply a measure of the degree by which u belongs to the set F. Here, it should be noted that the notation $\int$ represents only a fuzzy set and is not related to the usual integration or summation. There are several types of FMFs (Raol 2009). We employed the basic Gaussian-shaped function in our study, which is given by Eq. (5.59)

$$\mu_A(u) = \exp\left(-(u - a)^2/2b^2\right) \tag{5.59}$$

where a, b signifies the mean and standard deviation of the membership function respectively. The function is distributed around parameter a, and parameter b determines the width of the function.

5.7.1.3 Dempster-Shafer Theory of Evidence

As mentioned above, the evidence of the traditional probability theory is associated with only one possible event, whereas the evidence of the DS theory can be associated with multiple possible events, e.g., sets of events. The DS theory is advantageous owing to its ability to capture the natural behavior of reasoning by narrowing the hypothesis set down to a smaller number of possibilities as the evidence increases (Shafer 1976). It allows for the direct representation of the uncertainty in system responses, where an imprecise input can be characterized by a set or an interval, and the resulting output is a set or an interval.

The three important elements of the DS theory are the basic probability assignment (bpa), or m; the belief function, or Bel; and the plausibility function, or Pl. The bpa is a primitive form of evidence theory describing the degree of an event A with the frame of discernment Θ (commonly referred to as the power set), where m of the null set is 0, and the summation of the m degrees of all the subsets of Θ is 1. The value of the basic probability assignment for a given set A, represented as $m(A)$, pertains only to the set A and makes no additional claims about any subsets of A. On the other hand, any further evidence on the subsets of A would be represented by another value, i.e. $m(B)$ for $B \subset A$, which satisfies the condition

$$m : P(H) \rightarrow [0, 1] \tag{5.60}$$

$$m(\emptyset) = 0 \tag{5.61}$$

$$\sum_{A \in P(H)} m(A) = 1 \tag{5.62}$$

where $P(H)$ represents the power set of H, $\emptyset$ is the null set, and A is a set in the power set, i.e., $A \in P(H)$.

From the m values, the upper and lower bounds of an interval can be defined as equation (5.63). This interval contains the precise probability of a set of interest, and it is bounded by the belief and plausibility. The belief for a set A is defined as the sum of all m values of the proper subsets (B) of the set of interest (A) ($B \subseteq A$). The plausibility is the sum of all the m values of the sets (B) that intersect the set of interest (A) ($B \cap A \neq \emptyset$).

$$[Bel(A), Pl(A)] \tag{5.63}$$

$$Bel(A) = \sum_{B|B \subseteq A} m(B) \tag{5.64}$$

$$Pl(A) = \sum_{B|B \cap A \neq \emptyset} m(B) \tag{5.65}$$

Note that the sum of all the belief measures need not always be 1. Similarly, the sum of all the plausibility measures need not always be 1 because of their non additive property (Klir and Wierman 1998).

In addition to being derived from the basic probability assignment (m), the belief and plausibility measures can be derived from each other. The plausibility can be alternatively derived from the belief as follows:

$$Pl(A) = 1 - Bel(\overline{A}) \tag{5.66}$$

$$Bel(\overline{A}) = 1 - \sum_{B|B \cap A \neq \emptyset} m(B) \tag{5.67}$$

where $\overline{A}$ is the classical complement of A. This definition of plausibility in terms of belief is because all basic assignments must sum up to 1.

To obtain a combined probability assignment within the frame of discernment, Dempster's rule of evidence is applied. The combination of evidence defines the intersection of the subsets for the frame of discernment based on conjunctive pooled evidence. Consider an example involving two pieces of evidence B and C (m_1 and m_2); their combination can be expressed as

$$m_{12}(A) = \sum_{B \cap C} m_1(B)m_2(C)/(1 - K) \tag{5.68}$$

When $A \neq \emptyset, m_{12}(\emptyset) = 0$

$$K = \sum_{B \cap C} m_1(B)m_2(C) \tag{5.69}$$

where K represents the basic probability mass associated with conflict, which is determined by summing the products of the m values of all sets where the intersection is null. The result can be easily determined using the belief interval that has the maximum belief among all the subsets. After extracting confidence intervals for all potential composite events, DS theory allows for flexibility in the final decision-making criteria, which is highly dependent on the user's preferences and the specific application (Petrou, et al. 2014).

5.7.2   Original CANFET

The original CANFET is a framework with a high level of flexibility (as shown in Fig. 5.10). In this sense, it is adjustable in every step to develop a new CANFET-based classifier. Changing the decision process to a simpler technique leads to producing the Simplified CANFET. The classification, which is based on the calculation of the membership grade of each training sample, is the Direct-matching CANFET. The steps of the original CANFET are: (i) preprocessing and converting satellite data to Andrews' curves, (ii) extracting the TAD type of Andrews' curve, (iii) membership grade acquisition, and (iv) decision making by Dempster-Shafer theory of evidence. Moreover, we developed a visual-based technique to improve the classification accuracy called the "confidence zone". It takes an advantage of the visualization of the training curves, which is the determining factor of the non-overlapping areas in the plots of all training data. This technique is similar to human-perceived curve separations because it appears as the most discriminable area and gives more weight to the classification.

5.7.2.1   Andrews' Curve of Satellite Data

The Andrews' function (5.57) was applied to all image pixels as expressed by:

$$f_b(t) = \left(SWIR1/\sqrt{2} \right) + (NIR \cdot \sin(t)) + (Red \cdot \cos(t)) + (Green \cdot \sin(2t))$$
$$+(Blue \cdot \cos(2t))$$

$$(5.70)$$

where Blue, Green, Red, NIR, and SWIR1 are the values of bands 2 to 6 of Landsat 8 respectively.

However, Andrews' curves are strongly dependent on the order of the variables. Lower-frequency terms have a greater impact on the shape of the curve (Cesar and Fyfe, 2005). We plotted the set of curves having identical labels and manually compared them via visual interpretation. As mentioned previously, the shape of Andrews' curve is very sensitive to the order of variables, especially in lower frequency terms. Therefore, in (1) we plot them in different arrangements. For

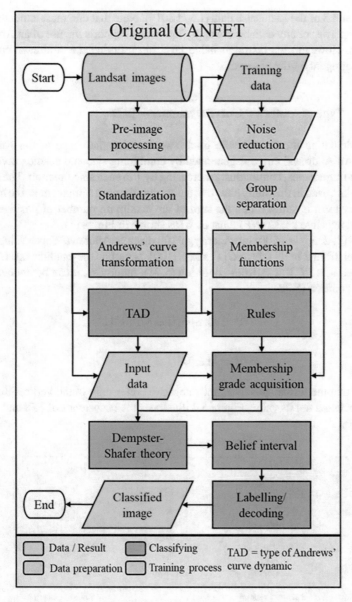

Fig. 5.10 The technical flow of the original CANFET

example, in the first round we plot the order of the bands as 2, 3, 4, 5, 6. However, class A and class B are not distinguishable. Plotting the order of the bands as 3, 4, 2, 5, 6, we also obtained no distinguishable results. We repeated this process until obtaining the appropriate order: 6, 5, 4, 3, 2. This order is supported by the spectral reflection of the plants, which lie mainly in the infrared and red bands, represented by bands 6 and 5 of the Landsat-8 data (Loyd 2014). Note that one must understand the basic properties of any data before applying, which supports the use of an Andrews' equation. However, the distinguishable level in observing the Andrews' curves is quite contingent on the observer.

5.7.2.2   Type of Andrews' Curve Dynamics—TAD

After calculating the membership grade, we examine the shape of the Andrews' curves. An Andrews' curve is generated by combining sine and cosine curves with different frequencies, continuously increasing by 1 in each sine function. The remote sensing data used in this study have five data points for each sample; thus, the highest-frequency term is $\sin(3t)$; i.e., the sum of the maximum number of positive peaks (PP) and negative peaks (NP) must be 6 (as shown in Fig. 5.11).

Let $G[1, 2, 3, \cdots , 100]$ be an array of 100 discrete Andrews' curve values and i be indices from 2 to 99 (since $G[1]$ and $G[100]$ cannot be the candidates). $G[i]$ can be defined as a PP if it satisfies condition (5.71); otherwise, it can be defined as an NP by condition (5.72).

$$G[i - 1] \leq G[i] \geq G[i + 1] \tag{5.71}$$

$$G[i - 1] \geq G[i] \leq G[i + 1] \tag{5.72}$$

Note that the terms "positive" and "negative" refer only to the vertical direction of the peak and not its value. Figure 5.4 illustrates the occurrence of PPs and NPs in an Andrews' curve.

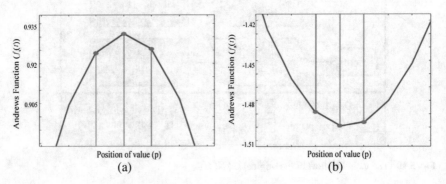

Fig. 5.11 Examples of (**a**) positive and (**b**) negative peaks from Andrews' curves

5.7.2.3 Confidence Zone and Weight

The confidence zone and weight are inspired by human recognition on a curve's analysis of plotting all training samples together. In a plot of all training samples, there are non-overlapping areas that can be used as a standard to separate one class from another. Figure 5.5 shows the confidence zones on Iris data that humans can easily distinguish.

For any position p in the plot of all training samples of a binary classification (class $C1$ and $C2$), the area determined as a confidence zone must satisfy the following conditions:

$$LL_{C1_p} > UL_{C2_p} \tag{5.73}$$

$$or \ LL_{C2_p} > UL_{C1_p} \tag{5.74}$$

where $UL = upper \ limit$, or the minimum value of the class at p, and $LL = lower \ limit$, or the maximum value of the class at p.

However, there are more than two categories in a classification. For this reason, we introduced a scoring method called confidence weight (CW). The method begins with calculating UL and LL for any position p of each training set. Table 5.9 shows the rules of the confidence weight scoring.

According to the Fig. 5.12, the weight score at position p of each x value can be assigned as shown in Table 5.10.

For each input, the sum of all CWs is divided by g (in this case, the maximum number of positions is 100). The final score can be used as an indicator, or a weight, in the decision process.

Table 5.9 The rules for the confidence weight scoring used in the study	Confidence levels	Scores	Conditions
	Very high	4	The input belongs to only one class, **Score given to that class**
	High	3	The input falls into at least two classes, **Score given to the classes**
	Medium	2	The input falls outside all classes, **Score given to all classes**
	Low	1	The input belongs to only one class, **Score given to the "other classes"**
	Very low	0	The input falls into at least two classes, **Score given to the "other classes"**

Fig. 5.12 The example position of an input value at p against four classes

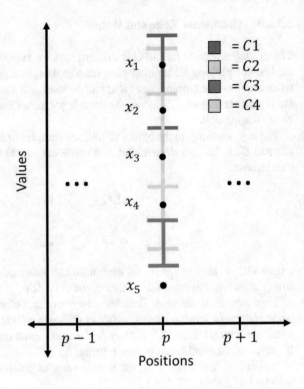

Table 5.10 The confidence weights from Fig. 5.12

Classes		C1	C2	C3	C4
x positions	x_1	3	0	0	3
	x_2	3	3	0	3
	x_3	0	3	0	3
	x_4	1	1	1	4
	x_5	2	2	2	2

5.7.2.4 Membership Grade Acquisition

The Andrews' curve of the data for each training set is obtained by projecting each data point onto the orthogonal function. For one set of training data, each continuous curve is fragmented into a set of 100 discrete values g, where the position p of value v has N points of training samples. Then, we calculate the mean a and standard deviation b as follows:

$$a_p = \frac{(v_1 + v_2 + v_3 + \cdots + v_N)}{N} \tag{5.75}$$

$$b_p = \sqrt{\frac{1}{N} \sum_{i=1}^{N} \left(v_i - a_p\right)^2} \qquad (5.76)$$

As per the Gaussian-shaped membership function (5.59), for any discrete input value at any position we can calculate the grade membership function in any training set A as follows:

$$\mu_A\left(v_p\right) = \exp\left(\frac{-\left(v_p - a_p\right)^2}{2b_p^2}\right) \qquad (5.77)$$

According to (5.72), only the matching position of the discrete value would be the input for each iterative Gaussian membership function. The illustration of Gaussian membership functions from a position of the discrete values is shown in Fig. 5.3d. The results are membership grades with regards to the positions of the discrete values. There were 100 values per pixel per membership function or subgroup. These fuzziness values represent the probability that each discrete input value belongs to each set of discrete values in the training set. Finally, we obtain the arithmetic sum of the values and divide by N to obtain only one value for each membership function, which is then compared with other values inside the class. We selected the candidate with the maximum membership value as a representative of the class.

5.7.2.5 Decision-Making by the Dempster-Shafer Theory of Evidence

The final goal is to determine class assignments based on belief and plausibility (confidence interval). Our workflow based on DS theory was based on beliefs and plausibility, as shown in Fig. 5.10. The results of fuzzification and the TAD were set as input for obtaining belief and plausibility. All input elements are normalized to get the basic probability distribution for each class. Based on their relationship with the basic probability distribution, the belief (5.64) and plausibility (5.65) equations are then employed. DS theory allows users to determine the final crisp classification (Petrou et al. 2014). We assign the class with the highest belief value to the output of the input curve pattern.

5.7.3 Simplified CANFET

The simplified CANFET is a compact version of the original CANFET. It does not have three things the original CANFET possesses: (i) DS-theory, (ii) TAD, and (iii) other decision rules. The decision process was replaced by using the maximum of the membership grade acquired from each class. Apart from fuzzy membership functions, there is no other statistical function applied to the Andrew curve of satellite data, which greatly reduces the complexity and computation time. Figure 5.13 shows the framework of simplified CANFET.

Therefore, it is easy to use, has very low complexity and very low, simple parameters, and is fast and easy to implement. However, its accuracy is not as high as that of other CANETs. Moreover, it seems impractical in some difficult classifications as it strongly depends on the training data.

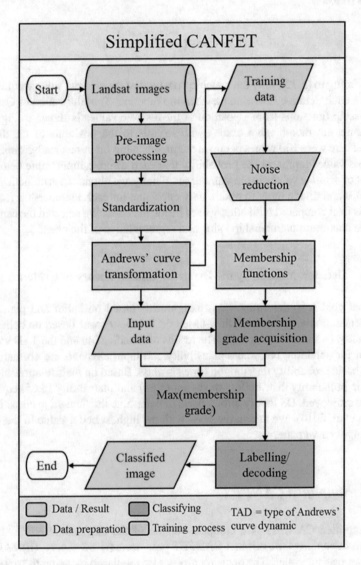

Fig. 5.13 The simplified CANFET framework

5.7.3.1 The Decision Rule for Simplified CANFET

The decision in the original CANFET method is based on DS theory. However, this method assumes that the data used in the research does not contain any defects or conflicts and does not use information from other sources; there exists only one piece of evidence and one event. Therefore, the calculation is simplified by finding the maximum arithmetic mean of the membership values in each group instead of the DS theory as shown in Eqs. (5.78) and (5.79). If the input has the highest mean membership value at C_N, then they will be assigned immediately to the class.

$$C_N = \frac{\sum_{i=1}^{n} m_i}{n} \tag{5.78}$$

$$Assigned class = \max(C_1, C_2, C_3, \cdots, C_N) \tag{5.79}$$

where C_N denotes class N and m is the membership grade of n discrete points.

5.7.4 Direct-Matching CANFET

Direct-matching CANFET was developed to solve the difficulty of training set separation in the original CANFET. The algorithm attempts to find the membership grade of the input against "all training samples" instead of "all training sets." Through this strategy, the system can use a smaller training scale if the user confidently trusts the training samples. Figure 5.14 shows the framework of Direct-matching CANFET.

Moreover, the noises in the training data will be ignored, as the input (labeled as a class, not noise sample) may obtain low membership grades from the noise data, because the input will match to a training curve with the most similar shape to the input. However, the direct-matching CANFET is not robust in terms of mislabeled training data and requires more computation than other CANFETs.

5.7.4.1 Direct-Matching Method

Contrary to the method of finding membership functions in the original CANFET, the Andrews' curve for *each training sample* is obtained by projecting each data point onto the orthogonal function. For the training sample, each continuous curve is segmented into a set of 100 discrete values g, where each position p has only one value v.

Per the Gaussian-shaped membership function (5.59), for any discrete input value at any position the grade membership function can be calculated in any sample A as follows:

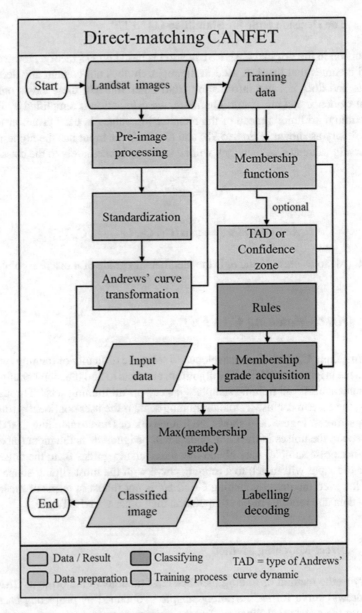

Fig. 5.14 The Direct-matching CANFET framework

$$\mu_A(v_p) = \exp\left(\frac{-(v_p - u_p)^2}{2b_p^2}\right) \tag{5.80}$$

where u_p is the discrete value of the training curve at position p, and b_p is directly assigned by the user (0.2 in this study).

According to (5.80), only the matching position of the discrete value would be the input of the Gaussian membership function for each iteration. The results are membership grades with regard to the discrete value locations. Each membership function has 100 values for each input value. These fuzziness values represent the probability that each discrete input value belongs to the discrete values of the training samples. Alternatively, the results may be weighted again by the TAD or confidence weight. Finally, we calculate the arithmetic sum of the values and divide by g to obtain only one value for a membership function and classify the inputs by the same decision of the simplified CANFET.

5.8   Summary

The fundamentals and mathematics for developing the enhanced classification method was explained in the first part of the chapter. In accordance with this knowledge and the relevant parameters, which were already mentioned in the previous chapter, the improved classification method was employed. The results show that there is a slight improvement in performance. However, it displays significant performs enhancement when applying a statistical test (z-scores). Compared to the iteration process of the Particle Filter, which is implemented for estimating the SVM parameters, the complexity of the proposed method is higher than the conventional one. However, it is regarded as a minor issue due to advances in computer technology.

References

Andrews D (1972) Plots of high-dimensional data. Biomatrics 28(1):125–136.
Arulampalam MS, Maskell S, Gordon N, Clapp T (2002) A tutorial on particle filters for online nonlinear/non-Gaussian Bayesian tracking. IEEE Transactions on Signal Processing 50(2):174–188
Boonprong S, Cao C, Torteeka P, Chen W (2017) A Novel Classification Technique of Landsat-8 OLI Image-Based Data Visualization: The Application of Andrews' Plots and Fuzzy Evidential Reasoning. Remote Sensing 9(5):427
Cesar GO, Fyfe C (2005) Visualization of high-dimensional data via orthogonal curves. Journal of Universal Computer Science 11(11):1806–1819
Dey A (2016) Machine Learning Algorithms: A Review. International Journal of Computer Science and Information Technologies 7(3):1174–1179
Gordon NJ, Salmond DJ, Smith AFM (1993) Novel approach to nonlinear/non-Gaussian Bayesian state estimation. Radar and Signal Processing, IEE Proceedings F 140(2):107–113

Grewal MS, Andrews AP (2001) Kalman filtering: theory and practice using MATLAB. Wiley, New York

Ho JD, Schon TB, Gustafsson F (2006) On resampling algorithms for particle filters. Available from: http://users.isy.liu.se/en/rt/schon/Publications/HolSG2006.pdf

Hsieh PF, Lee L, Chen NY (2001) Effect of spatial resolution on classification errors of pure and mixed pixels in remote sensing. IEEE Trans Geosci Remote Sens 39(12):2657–2663

Hsu C-W, Chang C-C, Lin C-J. 2008. A Practical Guide to Support Vector Classification. BJU International 101(1):1396–1400

Insom P, Chunxiang C, Boonsrimuang P, Di L, Saokarn A, Yomwan P, Yunfei X (2015) A Support Vector Machine-Based Particle Filter Method for Improved Flooding Classification. Geoscience and Remote Sensing Letters, IEEE 12(9):1943–1947

Jia YB (2015) Discrete-time kalman filter and the particle filter. Available from: http://web.cs.ias tate.edu/~cs577/handouts/kalman-filter.pdf

Johannes M, Polson N (2009) Particle Filtering. In: Mikosch T, Kreiß J-P, Davis RA, Andersen TG (eds) Handbook of Financial Time Series. Springer, Berlin Heidelberg, pp 1015–1029

Khushaba RN, Kodagoda S, Lal S, Dissanayake G (2011) Driver Drowsiness Classification Using Fuzzy Wavelet-Packet-Based Feature-Extraction Algorithm. Biomedical Engineering, IEEE Transactions on 58(1):121–131

Klir G, Wierman M (1998) Uncertainty based information: Elements of generalized information theory. Physica-Verlag GmbH & Co, Heidelberg, New York

Kotsiantis S (2007) Supervised machine learning: a review of classification techniques. Informatica 31:249–268

Lin C-J (2010) A Practical Guide to Support Vector Classification

Loyd C (2014) Landsat 8 bands « Landsat science. Retrieved Dec. 20, 2016, from http://landsat. gsfc.nasa.gov/landsat-8/landsat-8-bands

Petrou Z, Kosmidou V, Manakos I, Stathaki T, Adamo M, Tarantino C, Petrou M (2014) A rule-based classification methodology to handle uncertainty in habitat mapping employing evidential reasoning and fuzzy logic. Pattern Recogn Lett 48:24–33

Raol J (2009) Multi-sensor data fusion with MATLAB: Theory and practice. Taylor & Francis, Boca Raton

Ross SM. 2004. Introduction to Probability and Statistics for Engineers and Scientists. Elsevier Science

Salmond D, Gordon N (2005) An introduction to particle filters. Available from: http://dip.sun.ac. za/~herbst/MachineLearning/ExtraNotes/ParticleFilters.pdf

Schön TB. 2010. Solving Nonlinear State Estimation Problems Using Particle Filters - An Engineering Perspective

Shafer G (1976) A mathematical theory of evidence. Princeton University Press, Princeton, NJ, United States

Turner L (2013) An introduction to particle filtering. Available from: http://www.lancaster.ac.uk/ pg/turnerl/PartileFiltering.pdf

Yuan X, Sarma V (2011) Automatic Urban Water-Body Detection and Segmentation From Sparse ALSM Data via Spatially Constrained Model-Driven Clustering. Geoscience and Remote Sensing Letters, IEEE 8(1):73–77

Zarchan P, Musoff H (2009) Aeronautics AIo, Astronautics. Fundamentals of Kalman Filtering, A Practical Approach. American Institute of Aeronautics and Astronautics

Zhang Z (2011) Lecture notes 8: Nonparametric filters—the particle filter. Available from: http:// bcmi.sjtu.edu.cn/~zhzhang/papers/lec08.pdf

Part II
Waterborne Diseases Caused by Flooding Disasters

Chapter 6
Flood-Related Parameters Affecting Waterborne Diseases

Flooding is the leading cause of water-related morbidity and mortality in most monsoon areas frequently facing floods (Kazama et al. 2012). In particular, regions without clean water supplies and proper sewage systems that must depend on surface and sub-surface water polluted by inundation have the most serious problems with waterborne infectious diseases. There are 4 billion cases of diarrhea each year in addition to millions of other cases of illness associated with the lack of access to clean water (WHO 2000), and more than 5 million people die worldwide (Hunter 2003). About 75% of the infectious disease cases are reported in tropical areas, and about 50% of the deaths (4,800,000 people) occur in children under 5 years of age. Waterborne infectious disease takes hold in inundated areas during the flooding season. It is known that the flooded water distributes pollutants and contaminated matter from dumps (Smith 2001), and this commonly occurs in developing countries with increasing waterborne infectious diseases. Epidemics of diarrhea during flooding were reported in Sudan, India, and Mozambique in 1980, 1998, and 2000 respectively (WHO 2005). Schwartz et al. (2006) also reported clinical data for flood-associated diarrhea epidemics in Bangladesh. This chapter provides information on how floods relate to waterborne diseases, as well as relative parameters. Based on the flooding period which was already explained in Chap. 3 and the dataset in the Chap. 2, flood related parameters affecting waterborne diseases can be explained as follows in the following sections.

6.1 Flood Parameters Derived from Multi-temporal Remote Sensing Data

A series of flood maps in the study area was derived from the water extraction of the multi-temporal remote sensing data, including a series of Radarsat-2, HJ-1A/B and THEOS imageries. Flood areas at various times during September to December 2011

© Higher Education Press and Springer Nature Singapore Pte Ltd. 2021
C. Cao et al., *Environmental Remote Sensing in Flooding Areas*,
https://doi.org/10.1007/978-981-15-8202-8_6

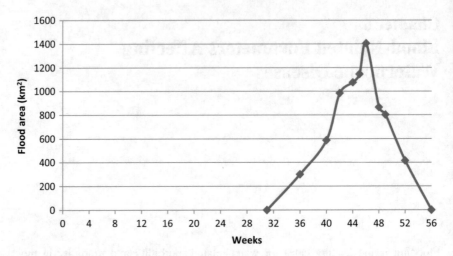

Fig. 6.1 The flood areas at various time in 2011 in the study area estimated from the water extraction of multi temporal remote sensing data

are important parameters, which relate to the probability of flood affected people to infect pathogen-contaminated floodwater in spatial analysis.

We obtained the flood duration in each area by comparing the maps of the flooded areas, which were derived from the multi-temporal remote sensing scenes. From Fig. 6.1, the resulting inundation areas were found by using multi temporal remote sensing data, which started approximately in week 36 with flood areas of about 300 km^2. Subsequently, the flood area gradually increased to 58,700 and 98,300 km^2 in weeks 40 and 42 respectively, until it peaked in week 46 with the inundated areas of 140,400 km^2. Then, flooding gradually decreased during weeks 49–52 with the flood area of 80,100 and 41,500 km^2 respectively.

Figure 6.2 provides the spatial distribution map of the flood duration. It shows that the inundated areas covered the entire study area at its peak. The longest inundated period was over four months, while the shortest was barely one week. As a first priority, the Thai government and private sectors aimed to prevent or reduce the water level in urban areas, along transportation routes and in residential and industrial areas by draining floodwater into agricultural areas, particularly rice fields, which are frequently located in low-lying areas. Therefore, the areas with the longest flood durations tend to be agricultural areas.

6.2 Population Variables

Outbreaks of disease are more frequent and more severe when the population density is high (WHO: Water Sanitation Health). For communities, inadequate shelter and

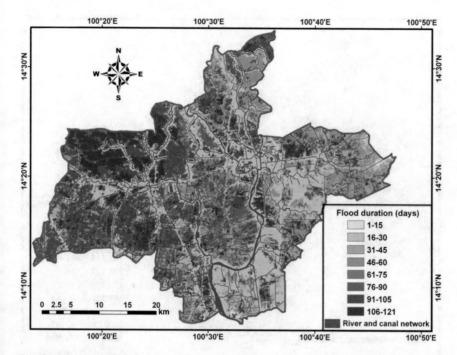

Fig. 6.2 Flood duration estimated from the multi-temporal remote sensing imagery during September to December 2011 in the study area

overcrowding are major factors in the transmission of diseases with epidemic potential, such as acute respiratory infections, meningitis, typhus, cholera, scabies, etc. Other public structures, such as health facilities, not only represent a concentrated area of patients, but also a concentrated area of germs. In an emergency, the number of hospital-associated infections will typically rise. Decreasing overcrowding by providing extra facilities and a proper organization of the sites or services in health-care facilities is a priority.

In this study, the census data of each community in the study area were utilized to calculate population density, as shown in Fig. 6.3.

Figure 6.3 illustrates the population density of the eight districts in terms of spatial distribution. From the figure, Phra Nakhon Si Ayudhya (simplified as Phra Nakhon in Fig. 6.3) has the highest population density, because it is the main district of Ayutthaya province. Bang Pa-in also has a high population density, because it is the location of large industrial estates, such as Bang Pa-in Industrial Estate and Hi-Tech Industrial Estate (Fig. 6.4), which have many thousands of employees. The population of other districts distributes along main roads and cluster in townships, at which are the locations of district hospitals as shown in Fig. 6.4.

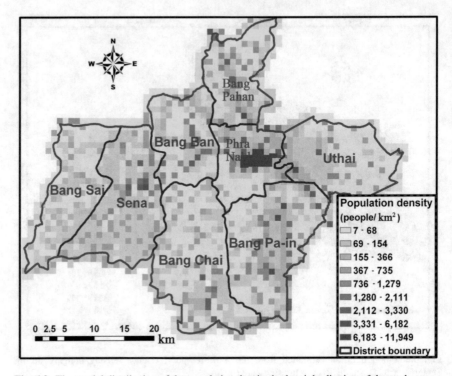

Fig. 6.3 The spatial distribution of the population density in the eight districts of the study area

6.3 Flood Water Quality

During flooding, pathogens from contaminated sources spread into inundation areas via water and sedimentation (Muirhead et al. 2004). Monitoring of the surface water should be aimed at achieving a representative quantification of the numbers of pathogenic microorganisms in the source water, in addition considering seasonal variability as well as short term fluctuations of pathogen concentrations (Westrell et al. 2006). In a flood disaster, fecal coliform bacteria, or *E. coli*, are common causes of diarrhea, because the flooding washes fecal material from human habitats, causing increased transmission of bacterial infection. To investigate waterborne disease, almost all studies have attempted to determine the presence of pathogens, such as bacteria and viruses, in contaminated water based on quantitative microbial risk assessment (QMRA). Time-consuming laboratory testing is required to determine the amount of pathogens in each water sample.

Dissolved oxygen (DO) is an important indicator of river health (the ecological condition of a river) and is used by regulators as part of the classification scheme for good chemical status (Williams and Boorman 2012). Therefore, researchers have frequently used DO to evaluate water quality (Kannel et al. 2007). A number of studies expressed a close relationship between DO and diarrheal pathogens, such

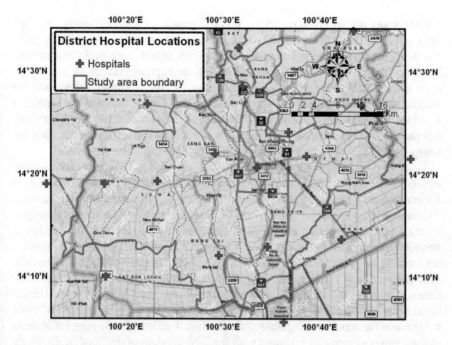

Fig. 6.4 The hospital locations distributed along main roads of each district in the study areas that relate to the spatial distribution of population density

as fecal coliform bacteria, *E. coli*, and *Vibrio cholerae* (Islam et al. 2007; Kersters et al. 1995; Massoud 2012; Osode and Okoh 2010). In particular, Osode and Okoh (2010) revealed that DO negatively correlated with *E. coli* densities ($P < 0.001$). The use of DO instead of parameters of the pathogens as an indicator of the water-quality surveillance system, as in many countries including Thailand (Boonsoong et al. 2010), can more comprehensively model the risk of waterborne disease for spatial analysis with up-to-date water-quality data. Therefore, this study used DO as an input factor governing the risk of diarrheal infection of people in inundated areas.

6.3.1   Inverse Distance Weighting for Spatial Distribution

For this study, we used the inverse distance weighting (IDW) function, one of the most frequently used deterministic models in spatial interpolation (Chen and Liu 2012; Srivastava et al. 2012a, b; Teegavarapu et al. 2012), to interpolate the spatial distribution of DO in the study area.

The IDW approach, a deterministic spatial interpolation model, is one of the most popular approaches developed by geoscientists and geographers and has been embedded in several GIS applications and packages. The main idea of this approach is that for any pair of given points the attribute values are related to each other, but their similarity is inversely related to the distance between the two locations. IDW has the advantage of being fast and easy to compute and quite straightforward to interpret. The general concept is based on the assumption that for an unsampled point the attribute value is the weighted average of nearby known values, and the weights are inversely dependent on the distances between the sampled and predicted locations. For our study, flooding covered the entire study area at its peak. Therefore, it is plausible to consider the study area as a continuous surface and to use IDW for interpolating the spatial distribution of DO samples in the entire study area.

Compared to geostatistical approaches such as kriging, IDW is a deterministic method (Bailey and Gatrell 1996). On the other hand, kriging is the preferred method for some spatial statistical analyses, but it is a relatively intricate and complex method. Although kriging has many types, almost all of them are based on similar concepts. Firstly, the data should be examined to determine the spatial structure, which is frequently expressed by the empirical variogram. Then after capturing the spatial autocorrelation structure of the data from the given empirical variogram, a mathematical equation is employed to fit the empirical variogram as the theoretical variogram function to model spatial autocorrelation. Consequently, weights are derived based upon this function. The combinations between linear and non-linear of these weights and the corresponding observed values are used to estimate values at unsampled locations or points. Compared with the IDW method as depicted in Eq. (6.1), Kriging requires extra steps to derive the empirical variogram and to fit a function to the variogram from which the weights are derived. An obvious advantage of Kriging is that Kriging variance is readily available during the computational process and it can serve as the confidence level of the interpolation, in addition to some other strengths (Zimmerman et al. 1999).

Otherwise, the assumption of IDW is very simple. The method assumes that Tobler's first law of geography (Tobler et al. 1970) is true for the data. However, there is no need to specify a theoretical distribution for the data. The method does not include computationally intensive procedures such as, for example, inverting the covariance matrix in kriging. A drawback of IDW is that sometimes the inverse distance weights are not determined by the empirical data of the study area. The provided adaptive IDW methodology is an attempt to update the weights with information derived from the empirical data.

The basic calculations for IDW (Lu and Wong 2008) are expressed in Eqs. (6.1) and (6.2). In Eq. (6.1), IDW estimates the unknown value $\hat{y}(S_0)$ in location S_0 given the observed y values for sampled locations S_i. The estimated value in S_0 is a linear combination between the weights (λ_i) and observed y values in S_i.

$$\hat{y}(S_0) = \sum_{i=1}^{n} \lambda_i y(S_i) \tag{6.1}$$

$$\lambda_i = d_{0i}^{-\alpha} / \sum_{i}^{n} d_{0i}^{-\alpha} \tag{6.2}$$

with

$$\sum_{i}^{n} \lambda_i = 1 \tag{6.3}$$

From (6.2), the numerator is the inverse of the distance (d_{0i}) between S_0 and S_i with a power α. The denominator is represented as the sum of all inverse-distance weights for all locations i so that the sum of all λ_i for an unsampled point will be unity. The α parameter is determined as a geometric form for the weight while other specifications are possible. This specification implies that if α is greater than 1, the so-called distance-decay effect will be greater than proportional to an increase in distance, and vice versa. Therefore, small α tends to yield computed values as means of S_i in the vicinity. Meanwhile large α tends to give higher weights to the nearest points and increasingly lower weights to points farther away. Thus, when $\alpha \to 0$ and $\lambda_i = 1/n$, then

$$y(S_0) = \sum_{i=1}^{n} \lambda_i y(S_i) = \sum_{i=1}^{n} \frac{1}{n} y(S_i) \tag{6.4}$$

Subsequently, the computed value is the mean of all sampled values. Let S_j be the nearest neighbour to S_0, and λ_i denote the distance between S_0 and S_i, then min $L_i= L_j$. When $\alpha \to \infty$, the weight will be defined in the following manner:

$$\lambda_i = \begin{cases} 1, & i = j\left(L_j = min\{L_i\}\right) \\ 0, & i \neq j \end{cases} \tag{6.5}$$

And

$$\hat{y}(S_0) = \sum_{i=1}^{n} \lambda_i y(S_i) = y\left(S_j\right) \tag{6.6}$$

In this case, the estimated value will be the same as the value in the nearest sampled location j.

To evaluate the interpolated results, cross-validation method is a widely used method. We therefore applied cross validation to evaluate the IDW interpolation for the spatial distribution of flood-related parameters used in this study.

6.3.2 Spatial Distribution of Dissolved Oxygen

In this study, DO was served as an input factor governing the risk of diarrheal infection of people in inundated areas. We obtained 186 DO samples from the Pollution Control Department of Thailand, which were taken during flooding, as shown in Fig. 6.5. As discussed above, we used IDW to interpolate these point measures into a continuous

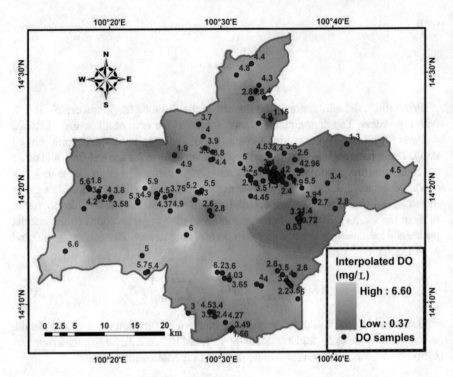

Fig. 6.5 The spatial distribution of dissolved oxygen (DO) estimated from 186 flood-water samples based on inverse distance weighting (IDW)

grid. The resulting map after interpolation, shown in Fig. 6.5, illustrates the DO spatial distribution. Higher DO values indicate better water quality. Thus, the redder areas in Fig. 6.5 have poorer water quality. Overlaying the water-quality map on a land-use layer, we found that the intensely red area on the bottom right of Fig. 6.5 is the location of major industrial estates in Bang Pa-in. The fecal and chemical materials leaking from industrial factories caused a very low value of DO in this area. The best water quality, with the highest values of DO, was mostly located in the countryside or in agricultural areas. The mean value of DO of each district is given in Table 6.1.

The root mean squared cross-validation error (Joseph and Kang 2011), which is an assessment of the uncertainty of the IDW interpolation of the DO spatial distribution, is 0.08 mg/L, while the resolutions of the DO meters, or the average standard errors of our DO measurements, range from 0.01 to 0.10 mg/L. The similarity between the interpolation error and the measurement error indicates that we are correctly assessing the variability in the interpolation.

Viewing all factors affecting the waterborne disease in this chapter, we calculated the mean values of each district as expressed in Table 6.1. Table 6.1 provides the means of three main parameters of each district in the study area. Being the main urban district area of Ayutthaya province, Phra Nakhon Si Ayudhya district has

Table 6.1 Means of the main parameter values affecting the waterborne disease used in this study divided by eight districts in the study area

District	Population	Area (km^2)	Pop. Density (people/km^2)	Mean flood duration (days)	Mean DO (mg/L)
Sena	66,122	215.28	307.14	67	3.6
Bang Ban	34,379	136.75	251.40	53	3.9
Uthai	46,540	170.62	272.77	40	3.3
Bang Sai	19,685	164.72	119.51	75	4.3
Bang Pa-in	90,188	237.10	380.38	35	3.0
Phra Nakhon Si Ayudhya	139,129	117.82	1180.86	30	3.6
Bang Chai	47,083	250.03	188.31	44	3.9
Bang Pahan	41,313	130.96	315.46	44	3.7

the highest mean value of population density (1180.86 people/km^2), followed by Bang Pa-in (380.38 people/km^2) and Bang Pahan (315.46 people/km^2), while the relatively long period of inundated areas covered the agriculture areas in the Bang Sai, Sena and Bang Ban district with the estimated mean flood duration of 75, 67, and 53 days respectively. Moreover, Bang Pa-in district, in which a number of major industrial estates are located, had the worst floodwater quality with a mean value DO of 3.0 mg/L.

Note that in this study almost DO samples were collected during October to November 2011 during the period in which the flood occurred. About 25% of DO samples were measured 2–5 times at various times. The average range between maximum and minimum DO values measured more than one time is 0.62 mg/L, while the significant changing value of DO for categorizing a class of water quality is approximately 1 mg/L (Massoud 2012). This implies that, for each sample point, a DO value, or its mean value collected during flooding, can be represented as the floodwater quality of that position. Thus, it is appropriate to estimate the outbreak risk due to floods by using the DO data collected during flooding, because of the close association with waterborne disease pathogens (Osode and Okoh 2010). However, due to the limitation of the temporal resolution of floodwater data used in the study, there is still further need of a study to detect the changing of floodwater DO value during flooding of each area, as well as its effect on the outbreak risk model.

6.4 Summary

The flood duration of each pixel was estimated by comparing a series of flood-classified map from multi-temporal remote sensing data. The flood duration map shows that the inundated areas covered the entire study area at its peak. The longest inundated period was over four months, while the shortest was barely one week.

We utilized the population density estimated from the census data as an important input parameter in this study, because the outbreaks are more frequent and more severe when the population density is high. Dissolved oxygen (DO) was used as the water quality indicator governing the risk of diarrheal infection of people in this study, because it closely relates to waterborne pathogens and is easy to measure. The inverse distance weighting (IDW) method, one of the most frequently used deterministic models in spatial interpolation, was employed to interpolate the spatial distribution of water samples and population data in the study area. For assessing the uncertainty of the interpolation, the root mean squared cross-validation error is 0.078 mg/L, while the resolutions of the DO meters, or the average standard errors of our DO measurements, range from 0.01 to 0.10 mg/L. The interpolation error being in the range of measurement error indicates the acceptable assessment of the interpolation.

References

Bailey TC, Gatrell AC (1996) Interactive spatial data analysis (vol 22, pg 272, 1996). Environ Int 22(5):656–656

Boonsoong B, Sangpradub N, Barbour MT, Simachaya W (2010) An implementation plan for using biological indicators to improve assessment of water quality in Thailand. Environ Monit Assess 165(1–4):205–215

Chen FW, Liu CW (2012) Estimation of the spatial rainfall distribution using inverse distance weighting (IDW) in the middle of Taiwan. Paddy Water Environ 10(3):209–222

Hunter PR (2003) National disease burden due to waterborne transmission of nosocomial pathogens is substantially overestimated. Arch Intern Med 163(16):1974

Islam MS, Brooks A, Kabir MS, Jahid IK, Islam MS, Goswami D, Nair GB, Larson C, Yukiko W, Luby S (2007) Faecal contamination of drinking water sources of Dhaka city during the 2004 flood in Bangladesh and use of disinfectants for water treatment. J Appl Microbiol 103(1):80–87

Joseph VR, Kang LL (2011) Regression-based inverse distance weighting with applications to computer experiments. Technometrics 53(3):254–265

Kannel PR, Lee S, Lee YS, Kanel SR, Khan SP (2007) Application of water quality indices and dissolved oxygen as indicators for river water classification and urban impact assessment. Environ Monit Assess 132(1–3):93–110

Kazama S, Aizawa T, Watanabe T, Ranjan P, Gunawardhana L, Amano A (2012) A quantitative risk assessment of waterborne infectious disease in the inundation area of a tropical monsoon region. Sustain Sci 7(1):45–54

Kersters I, Vanvooren L, Huys G, Janssen P, Kersters K, Verstraete W (1995) Influence of temperature and process technology on the occurrence of aeromonas species and hygienic indicator organisms in drinking-water production plants. Microb Ecol 30(2):203–218

Lu GY, Wong DW (2008) An adaptive inverse-distance weighting spatial interpolation technique. Comput Geosci 34(9):1044–1055

Massoud MA (2012) Assessment of water quality along a recreational section of the Damour River in Lebanon using the water quality index. Environ Monit Assess 184(7):4151–4160

Muirhead RW, Davies-Colley RJ, Donnison AM, Nagels JW (2004) Faecal bacteria yields in artificial flood events: quantifying in-stream stores. Water Res 38(5):1215–1224

Osode AN, Okoh AI (2010) Survival of free-living and plankton-associated Escherichia coli in the final effluents of a waste water treatment facility in a peri-urban community of the Eastern Cape Province of South Africa. Afr J Microbiol Res 4(13):1424–1432

Schwartz BS, Harris JB, Khan AI, Larocque RC, Sack DA, Malek MA, Faruque ASG, Qadri F, Calderwood SB, Luby SP et al (2006) Diarrheal epidemics in Dhaka, Bangladesh, during three consecutive floods: 1988, 1998, and 2004. Am J Trop Med Hyg 74(6):1067–1073

Smith E (2001) Pollutant concentrations of stormwater and captured sediment in flood control sumps draining an urban watershed. Water Res 35(13):3117–3126

Srivastava P, Gupta M, Mukherjee S (2012a) Mapping spatial distribution of pollutants in groundwater of a tropical area of India using remote sensing and GIS. Appl Geomat 4(1):21–32

Srivastava PK, Han DW, Gupta M, Mukherjee S (2012b) Integrated framework for monitoring groundwater pollution using a geographical information system and multivariate analysis. Hydrol Sci J-J Des S Hydrol 57(7):1453–1472

Teegavarapu RSV, Meskele T, Pathak CS (2012) Geo-spatial grid-based transformations of precipitation estimates using spatial interpolation methods. Comput Geosci 40:28–39

Tobler WR, Mielke HW, Detwyler TR (1970) Geobotanical distance between New Zealand and neighboring islands. Biosci 20(9):537–542

Westrell T, Teunis P, van den Berg H, Lodder W, Ketelaars H, Stenstrom TA, Husman AMD (2006) Short- and long-term variations of norovirus concentrations in the Meuse river during a 2-year study period. Water Res 40(14):2613–2620

Williams RJ, Boorman DB (2012) Modelling in-stream temperature and dissolved oxygen at sub-daily time steps: An application to the River Kennet, UK. Sci Total Environ 423:104–110

WHO (2000) The World Health Report 2000. Health systems, improving performance, geneva, world health organization

WHO (2005) Social health insurance, selected case studies from Asia and the Pacific. WHO Library Cataloguing in Publication Data. 34–60

Zimmerman D, Pavlik C, Ruggles A, Armstrong MP (1999) An experimental comparison of ordinary and universal kriging and inverse distance weighting. Math Geol 31(4):375–390

Chapter 7
Measure of Disease Risk

7.1 Waterborne Diseases Caused by Flooding

Floods cause an increase of pathogen concentrations in natural waters (Funari et al. 2012) that results from over-flooding of the runoff of animal manure, sewage treatment plants, and the remobilization and redistribution of contaminated sediments(Hilscherova et al. 2007; Nagels et al. 2002), as shown in Fig. 7.1.

Because the distribution of pathogens relies on the hydrodynamics of surface water bodies, it can be expected that speeding up water fluxes carrying pathogens through flooding and heavy rainfall will counteract the natural pathogen activation by temperature and UV in the environment. Enhanced environmental levels of pathogenic microorganisms may lead to increasing incidence of diseases and occurrence of new ones (Boxall et al. 2009). Generally, it is expected that zoonotic infections may spread due to an increased washing into water of livestock and wild animal feces. In addition, viruses can be the unseen etiological pathogens responsible for human diseases even when waters meet regulatory criteria for fecal contamination based on conventional bacterial indicators that are less defiant than viruses and decay much faster in a natural environment (Maalouf et al. 2010). A number of reports reveal the association between waterborne disease outbreaks and flood disasters or excessive rainfall (Funari et al. 2012). For example, because of heavy rainfall, associated runoff, and consequent contamination of Milwaukee Lake, in Milwaukee 1993 the largest reported waterborne disease outbreak in the United States, due to the presence of Cryptosporidium cists in drinking water, caused the deaths of 54 people and the hospitalization of more than 403,000 patients (Hoxie et al. 1997; Mackenzie et al. 1994). In the European Union, in 2007, seventeen waterborne outbreaks were reported by eight countries, and under-reporting indicated the morbidity of 10,912 cases with 232 hospitalizations.

Furthermore, the impact of floods and storms disrupting the water distribution system by mixing drinking and wastewaters also have a significant effect on the diffusion of cholera caused by the naturally occurring *Vibrio cholerae*. The disease

© Higher Education Press and Springer Nature Singapore Pte Ltd. 2021
C. Cao et al., *Environmental Remote Sensing in Flooding Areas*,
https://doi.org/10.1007/978-981-15-8202-8_7

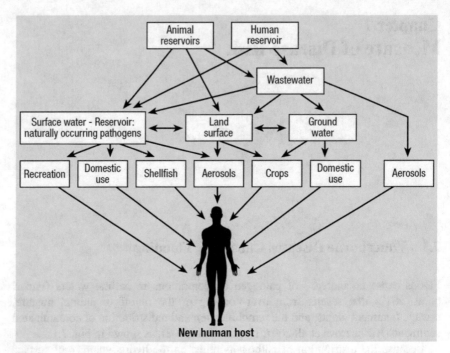

Fig. 7.1 Sources of waterborne pathogen related to floods that are naturally present in surface waters, such as a reservoir (modified from Funari et al. (2012))

is one of the most severe forms of waterborne diarrheal disease, especially for developing countries where outbreaks occur seasonally and are closely related to poverty and use of poor sanitation and unsafe water. Extreme climatic events also cause an increase in cases of diseases and fatalities by adding to an oral-fecal contamination pathway that is difficult to manage.

In this study, we attempt to find the relationship between the increasing rate of three waterborne diseases and the temporal pattern change of inundation areas during flooding. The three waterborne diseases in the study are (1) diarrhea, (2) conjunctivitis and (3) leptospirosis.

1. Diarrhea

Diarrhea, i.e. diarrhoea, is the passage of three or more loose or liquid stools per day, or more frequently than is normal for the individual. It is usually a symptom of gastrointestinal infection, which can be caused by a variety of bacterial, viral and parasitic organisms. Infection is spread through contaminated food or drinking water, or from person to person because of poor hygiene. Severe diarrhoea leads to fluid loss, and may be life threatening, particularly in young children and people who are malnourished or have impaired immunity.

A number of papers have revealed the association between floods and diarrheal diseases (Ding et al. 2013). An increase in diarrheal diseases in post-flood periods has been reported in the 1988 Bangladesh and Khartoum floods (Shears 1988; Siddique et al. 1991). The main risk factors of the diarrhea epidemic in the 1998 Bangladesh floods included lack of distribution of water purification tablets and the type of water storage vessels (Kunii et al. 2002). In China, Ding et al. (2013) found that floods have significantly increased the risks of infectious diarrhea by persistent and heavy rainfall in the upper and middle Huaihe River of China in 2007. In high-income countries, the risk of diarrheal illness appears to be lower during flooding than that of developing countries. A survey from Germany found that the prevalence of diarrhea was 6.9% in flooded areas with the main risk factors being contact with floodwater, women, and water supply from a private pond. In Thailand, the related public health agencies received several reports regarding diarrheal outbreaks during floods. However, very little research has been conducted to investigate their correlation.

2. Conjunctivitis

Conjunctivitis is a common eye condition worldwide. It causes inflammation (swelling) of the conjunctiva: the thin layer that lines the inside of the eyelid and covers the white part of the eye. Conjunctivitis is often called "pink eye" or "red eye" because it can cause the white of the eye to take on a pink or red coloration. The most common causes of conjunctivitis are viruses, bacteria, and allergens. However, there are other causes, including chemicals, fungi, certain diseases, and contact lens wear (especially wearing lenses overnight). The conjunctiva can also become irritated by foreign bodies in the eye and by indoor and outdoor air pollution caused, for example, by chemical vapors, fumes, smoke, or dust.

Conjunctivitis caused by viruses, known as Acute Hemorrhagic Conjunctivitis (AHC), is a highly communicable disease in tropical countries as well as in Thailand. People in Thailand frequently face epidemics of conjunctivitis almost every year, especially during the rainy season that causes wet conditions in which the virus can grow easily.

3. Leptospirosis

Leptospiros is a bacterial disease that affects both humans and animals. Humans become infected through direct contact with the urine of infected animals or with a urine-contaminated environment. The bacteria enter the body through cuts or abrasions on the skin, or through the mucous membranes of the mouth, nose and eyes. Person-to-person transmission is rare. In the early stages of the disease, symptoms include high fever, severe headache, muscle pain, chills, redness of the eyes, abdominal pain, jaundice, hemorrhaging in the skin and mucous membranes, vomiting, diarrhea, and rash.

The annual number of estimated severe leptospirosis cases is around 500,000 worldwide, with an average fatality rate approaching 10% (Hartskeerl et al. 2011). The incidence of this disease is higher in tropical regions with low socioeconomic conditions and where there is environmental proximity to infected animals (Bharti et al. 2003; Pappas et al. 2008; Reis et al. 2008). However, recently, leptospirosis has become an urban health problem not only in developing countries, but also in industrialized countries, such as Germany, Japan, and the USA (Jansen et al. 2005; Koizumi et al. 2009; Vinetz et al. 1996).

Leptospirosis is an endemic infection throughout the tropics, where outbreaks are well-described, often in the context of heavy freshwater flooding (Amilasan et al. 2012; Trevejo et al. 1998). Regarding the 2011 Thailand major flood, the Bureau of Epidemiology of Thailand monitored flooding of the Bangkok Metropolitan Region between late October of 2011 and January of 2012 and found that the flooding posed a potential risk for a leptospirosis outbreak in case reports. Thaipadungpanit et al. (2013) provided baseline information on environmental leptospira in Bangkok together with a set of laboratory tests that could be readily deployed in the event of future flooding.

7.2   Disease Risk Assessment

The risk of contracting an illness can be expressed as the probability of infection or illness during a defined period or may be attributed to an exposure. Analytical studies can provide a direct estimate of individual risk, and the incidence of illness among the unexposed and exposed can be directly compared. The basic measures generally used are the risk or rate difference, incidence rate ratio (IRR), cumulative incidence ratio, or odds ratio (Craun et al. 2006). To determine IRR of diarrhea due to flooding, we adapted the IRR equation from the CDC (2012) as expressed in Eq. (7.1). We defined the period as the flood duration and designated a constant equal to 5, which transforms the result of the division into a uniform quantity (n), for fitting with the number of patients in our study.

$$\text{IRR} = \frac{\text{New cases occurring during a given time period}}{\text{Size of population during the same time period}} \times 10^n \qquad (7.1)$$

From the morbidity datasets as shown in Sect. 2.3, we calculated the weekly IRR of leptospirosis, conjunctivitis, and diarrhea from Eq (7.1). The medians of each district are provided in Table 7.1 and Fig. 7.2.

Table 7.1 and Fig. 7.2 illustrates that in 2011 the median weekly IRR of diarrhea has the highest value followed by conjunctivitis and leptospirosis respectively. The three districts having the highest median values of diarrhea are Phra Nakhon Si Ayudhya (74.39 cases/10^5 population), Uthai (53.72 cases/10^5 population) and Sena (52.18 cases/10^5 population), while for conjunctivitis they are Phra Nakhon Si Ayudhya (8.63 cases/10^5 population), Bang Pa-in (7.76 cases/10^5 population) and Bang Sai (7.62 cases/10^5 population). Only three districts had leptospirosis patients in 2011: Uthai (0.17 cases/10^5 population), Bang Pa-in (0.11 cases/10^5 population) and Phra Nakhon Si Ayudhya (0.01 cases/10^5 population).

Table 7.1 The 2011 medians weekly incidence rate ratio (IRR) of leptospirosis, conjunctivitis, and diarrhea of the eight districts in the study area

District	Median weekly IRR in 2011 (cases/10⁵ population)		
	Leptospirosis	Conjunctivitis	Diarrhea
Phra Nakhon Si Ayudhya	0.01	8.63	74.39
Bang Chai	0	6.37	44.60
Bang Ban	0	5.82	26.18
Bang Pa-in	0.11	7.76	48.79
Bang Pahan	0	2.42	41.15
Sena	0	7.56	52.18
Bang Sai	0	7.62	40.64
Uthai	0.17	6.45	53.72
Entire Study Area	**0.04**	**6.09**	**50.37**

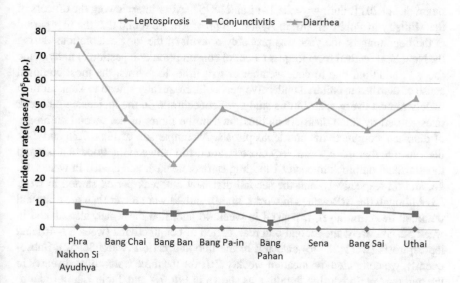

Fig. 7.2 Comparison of the 2011 medians weekly incidence rate ratio (IRR) of leptospirosis, conjunctivitis, and diarrhea divided by the eight districts in the study area

7.3   Outbreak Detection Methods

In this study, we want to detect the outbreak periods of leptospirosis, conjunctivitis, and diarrhea. However, there is no generally accepted number or percentage increase of cases that defines an outbreak or epidemic (Craun et al. 2006). What public health officials consider an outbreak of epidemic is usually based on previous surveillance, with an outbreak identified as the number of cases being greater than expected for a specific disease or set of symptoms in that area. The IRR of an outbreak period is expected to be higher than that of other periods in the same area. For this study, we set the definition of an outbreak (Schwartz et al. 2006) after plotting the association between the weekly IRR of diarrhea, conjunctivitis and leptospirosis and the flooded areas as derived from multi-temporal remote sensing data. Subsequently, the outbreak periods were defined when the weekly IRR exceeded the 2011 weekly IRR medians of each disease in the study area. Therefore, the outbreak detection levels, i.e. the 2011 weekly IRR medians of diarrhea, conjunctivitis and leptospirosis are 50.37, 6.09 and 0.04 cases/10^5 population respectively.

The outbreak periods of each disease and its relationship with the inundated areas derived from multi-temporal remote sensing data are provided in Fig. 7.3. Figure 7.3a provides flood areas at various times from temporal remote sensing data from weeks 36–52. This flood period time has a very close relationship with the outbreak period 2 of diarrhea, conjunctivitis and leptospirosis as shown in Fig. 7.3a, b, and c respectively.

From Fig. 7.3b, we found that the study area had two main periods of diarrheal outbreaks in 2011, during weeks 1–11 and 38–51. After interviewing the officers of the Ministry of Public Health in Ayutthaya province, we found that the first period at the beginning of the year was probably a result of the long celebrations during the New Year festival from frequent alcohol consumption or ingestion of unhygienic food. In addition, due to cool weather at that time, Rotavirus, the most common cause of diarrhea in children under five years old, frequently spread to Thai children widely almost every year. In this study, we concentrate on the increasing outbreak rates caused by flood disasters as illustrated in the figure of the second outbreaks of diarrhea, conjunctivitis and leptospirosis. A comparison of Figs. 7.3a and 7.3b demonstrate that there was a very close relationship between the flooded areas and the diarrheal outbreak intensity. Flooding started, peaked, and abated in weeks 36, 46, and 52 respectively, and the second diarrheal outbreak period started in week 38. Although the IRR values in weeks 39, 40, and 43 were lower than the diarrheal outbreak detection level, it peaked in week 46 and, like the water, also abated in week 52. Similarly, the second outbreak period of conjunctivitis (week 37–46) and leptospirosis (week 38–46) also fell in the flooding period (week 36–52). Subsequently, we computed the mean of weekly IRR for the three waterborne diseases in the outbreak period during flooding, as shown in Fig. 7.4 and Table 7.2, in order to estimate the outbreak risk. It shows that Phra Nakhon Si Ayudhya had the maximum mean value of both conjunctivitis and diarrhea while Bang Pahan had the minimum mean value of both conjunctivitis and diarrhea.

7.4 Estimation of Outbreak Risk Using Risk Ratio Function

In comparing the rates of disease of two groups, the relative risk, or risk ratio (RR), is widely used to compare the disease rates that differ by demographic characteristics or exposure histories as shown in Eq. (7.2) (CDC 2012; Craun et al. 2006). In this

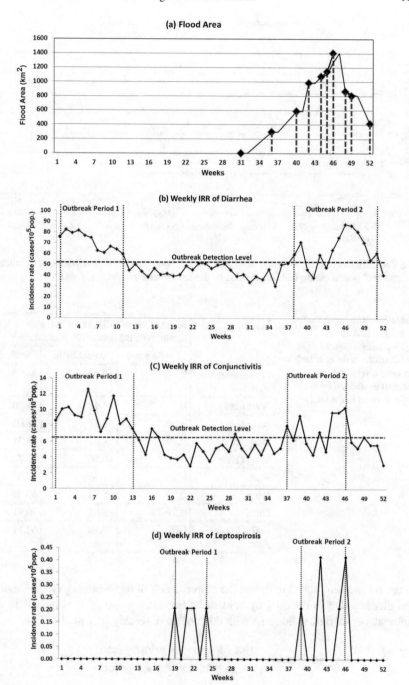

Fig. 7.3 The association between (**a**) changing of inundated areas during week 36–52 and defining of outbreak periods of (**b**) diarrhea, (**c**) conjunctivitis and (**d**) leptospirosis

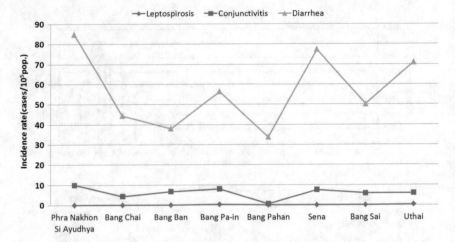

Fig. 7.4 Pattern of the mean values of weekly IRR of leptospirosis, conjunctivitis and diarrhea in the outbreak period during flooding divided by the eight districts in the study area

Table 7.2 The mean values of weekly IRR of leptospirosis, conjunctivitis and diarrhea in the outbreak period during flood occurrence divided by the eight districts in the study area

District	Mean of weekly IRR in the outbreak period during flooding (cases/10^5 populations)		
	Leptospirosis	Conjunctivitis	Diarrhea
Phra Nakhon Si Ayudhya	0	10.01	84.76
Bang Chai	0	4.41	44.30
Bang Ban	0	6.71	38.02
Bang Pa-in	0.32	8.02	56.31
Bang Pahan	0	0.56	33.71
Sena	0	7.56	77.35
Bang Sai	0	5.86	50.07
Uthai	0.31	5.95	70.91
Entire Study Area	0.09	7.16	64.11

study, RR was employed to define the outbreak risk of leptospirosis, conjunctivitis, and diarrhea due to flooding by comparing the mean value of weekly IRR in the outbreak period during flooding with the median of weekly IRR in 2011.

$$\text{Risk ratio} = \frac{\text{risk for group of primary interest}}{\text{risk for comparison group}} \qquad (7.2)$$

To estimate the outbreak risk due to flooding for each district, the RR was used to compare the IRR of weekly morbidity with the outbreak detection level or the median of weekly morbidity. An RR less than or equal to 1 indicates no outbreak or no risk. Otherwise it indicates the possibility of an outbreak, with outbreak intensity increasing with larger values of RR (Craun et al. 2006). The computed risk ratios of leptospirosis, conjunctivitis and diarrhea divided by the eight districts are shown in Fig. 7.5 and Table 7.3.

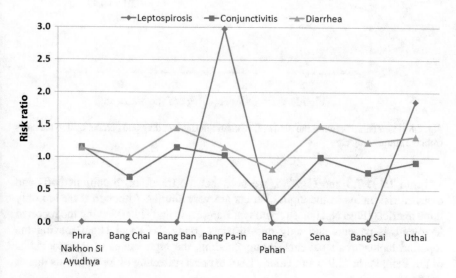

Fig. 7.5 Plotting of the outbreak risk ratios of the three waterborne diseases divided by the eight districts in the study area during the 2011 flood

Table 7.3 The risk ratios of leptospirosis, conjunctivitis and diarrhea during flooding divided by the eight districts in the study area

District	Risk ratio (RR)		
	Leptospirosis	Conjunctivitis	Diarrhea
Phra Nakhon Si Ayudhya	0	1.16	1.14
Bang Chai	0	0.69	0.99
Bang Ban	0	1.15	1.45
Bang Pa-in	2.97	1.03	1.15
Bang Pahan	0	0.23	0.82
Sena	0	1.00	1.48
Bang Sai	0	0.77	1.23
Uthai	1.86	0.92	1.32
Entire Study Area	2.23	1.18	1.27

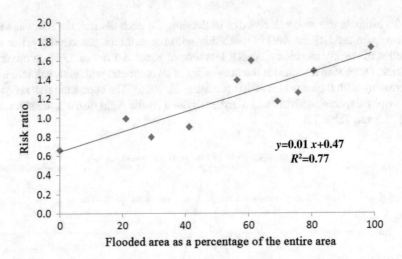

Fig. 7.6 Correlation between the diarrheal risk ratio during flooding and percent of flooded areas from the entire study area

From Table 7.3 and Fig. 7.5, we note that the trends of th conjunctivitis and diarrhea risk ratios in the eight districts are very similar. Although there are only three districts (Phra Nakhon Si Ayudhya, Bang Pa-in and Uthai) having the reported leptospirosis patients and only two districts (Bang Pa-in and Uthai) having the reported leptospirosis patients during flooding, the high values of the risk ratios of the Bang Pa-in (2.97) and Uthai (1.86) express spreading of leptospirosis due to the flood.

To determine the correlation between the outbreak risks and the flood occurrence, we plotted graphs between diarrheal risk ratios and conjunctivitis risk ratios with flood areas as shown in Fig. 7.6 and Fig. 7.7 respectively.

Figures 7.6 and 7.7 reveal the very close relationship between the risk ratios of diarrhea and conjunctivitis and the flood-inundated areas. The coefficient of determination (R^2) between the diarrheal risk ratios and flooded areas is 0.77, while the R^2 between the conjunctivitis risk ratios and flooded areas is 0.76. These results affirm that flood disasters lead to the outbreaks of waterborne diseases considerably.

7.5 Summary

We adapted the incidence rate ratio (IRR) equation, which can express the probability of infection during a defined period, to analyze the increasing of the outbreak rate of three waterborne diseases due to flooding. The 2011 weekly IRR median and mean of diarrhea, conjunctivitis and leptospirosis are 50.37 and 6.09 with 0.04 patients/100,000 populations. The outbreak periods of the waterborne diseases were defined from considering the plotting of the correlation between the weekly IRR and

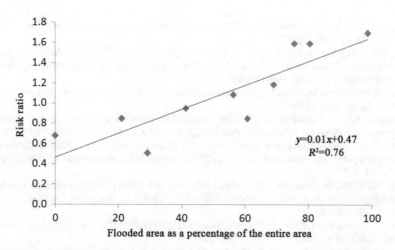

Fig. 7.7 Correlation between the conjunctivitis risk ratio during flooding and percent of flooded areas from the entire study area

the inundated areas. The weekly IRRs exceeded the outbreak detection levels; i.e. the medians of the 2011 weekly IRRs were determined to be exceedingly high during the outbreak periods. For the whole second outbreak period in 2011 diarrhea (week 38–51), conjunctivitis (week 37–46) and leptospirosis (week 38–46) spread in the flooded period (week 36–52). The risk ratio (RR) equation was employed to define the outbreak risk of waterborne diseases due to flooding by comparing the mean of weekly IRRs during flooding and the IRRs at the outbreak detection level, which are the medians of the 2011 weekly IRRs for this study. The outbreak risk ratios of the three waterborne diseases have a very close relationship with the inundated areas. The coefficient of determination (R^2) of the diarrheal risk ratio and the conjunctivitis risk ratio compared with the flooded areas are 0.77 and 0.76 respectively. These results affirm that flood disasters lead to the outbreaks of waterborne diseases considerably.

References

Amilasan AST, Ujiie M, Suzuki M, Salva E, Belo MCP, Koizumi N, Yoshimatsu K, Schmidt WP, Marte S, Dimaano EM (2012) Outbreak of leptospirosis after flood, the Philippines, 2009. Emerg Infect Dis 18(1):91–94

Bharti AR, Nally JE, Ricaldi JN, Matthias MA, Diaz MM, Lovett MA, Levett PN, Gilman RH, Willig MR, Gotuzzo E (2003) Leptospirosis: a zoonotic disease of global importance. Lancet Infect Dis 3(12):757–771

Boxall ABA, Hardy A, Beulke S, Boucard T, Burgin L, Falloon PD, Haygarth PM, Hutchinson T, Kovats RS, Leonardi G et al (2009) Impacts of climate change on indirect human exposure to pathogens and chemicals from agriculture. Environ Health Perspect 117(4):508–514

CDC (2012) Principles of epidemiology in public health practice: an introduction to applied epidemiology and biostatistics. CDC

Craun GF, Calderon RL, Wade TJ (2006) Assessing waterborne risks: an introduction. J Water
    Health 4(Suppl 2):3–18
Ding GY, Zhang Y, Gao L, Ma W, Li XJ, Liu J, Liu QY, Jiang BF (2013) Quantitative analysis of
    burden of infectious diarrhea associated with floods in Northwest of Anhui Province, China: a
    mixed method evaluation. Plos One 8(6):e65112
Funari E, Manganelli M, Sinisi L (2012) Impact of climate change on waterborne diseases. Ann Ist
    Super Sanita 48(4):473–487
Hartskeerl RA, Collares-Pereira M, Ellis WA (2011) Emergence, control and re-emerging
    leptospirosis: dynamics of infection in the changing world. Clin Microbiol Infect 17(4):494–501
Hilscherova K, Dusek L, Kubik V, Cupr P, Hofman J, Klanova J, Holoubek I (2007) Redistribution
    of organic pollutants in river sediments and alluvial soils related to major floods. J Soils Sediments
    7(3):167–177
Hoxie NJ, Davis JP, Vergeront JM, Nashold RD, Blair KA (1997) Cryptosporidiosis-associated
    mortality following a massive waterborne outbreak in Milwaukee, Wisconsin. Am J Public Health
    87(12):2032–2035
Jansen A, Schoneberg I, Frank C, Alpers K, Schneider T, Stark K (2005) Leptospirosis in Germany,
    1962–2003. Emerg Infect Dis 11(7):1048–1054
Koizumi N, Muto M, Tanikawa T, Mizutani H, Sohmura Y, Hayashi E, Akao N, Hoshino M,
    Kawabata H, Watanabe H (2009) Human leptospirosis cases and the prevalence of rats harbouring
    Leptospira interrogans in urban areas of Tokyo, Japan. J Med Microbiol 58(9):1227–1230
Kunii O, Nakamura S, Abdur R, Wakai S (2002) The impact on health and risk factors of the
    diarrhoea epidemics in the 1998 Bangladesh floods. Public Health 116(2):68–74
Maalouf H, Pommepuy M, Le Guyader FS (2010) Environmental conditions leading to shellfish
    contamination and related outbreaks. Food Environ Virol 2(3):136–145
Mackenzie WR, Hoxie NJ, Proctor ME, Gradus MS, Blair KA, Peterson DE, Kazmierczak JJ,
    Addiss DG, Fox KR, Rose JB et al (1994) A Massive outbreak in Milwaukee of cryptosporidium
    infection transmitted through the public water-supply. N Engl J Med 331(3):161–167
Nagels JW, Davies-Colley RJ, Donnison AM, Muirhead RW (2002) Faecal contamination over
    flood events in a pastoral agricultural stream in New Zealand. Water Sci Technol 45(12):45–52
Pappas G, Papadimitriou P, Siozopoulou V, Christou L, Akritidis N (2008) The globalization of
    leptospirosis: worldwide incidence trends. Int J Infect Dis 12(4):351–357
Reis RB, Ribeiro GS, Felzemburgh RDM, Santana FS, Mohr S, Melendez AXTO, Queiroz A,
    Santos AC, Ravines RR, Tassinari WS et al (2008) Impact of environment and social gradient on
    leptospira infection in urban slums. Plos Negl Trop Dis 2(4):e228
Schwartz BS, Harris JB, Khan AI, Larocque RC, Sack DA, Malek MA, Faruque ASG, Qadri F,
    Calderwood SB, Luby SP et al (2006) Diarrheal epidemics in Dhaka, Bangladesh, during three
    consecutive floods: 1988, 1998, and 2004. Am J Trop Med Hyg 74(6):1067–1073
Shears P (1988) The Khartoum floods and diarrheal diseases. Lancet 2(8609):517–517
Siddique AK, Baqui AH, Eusof A, Zaman K (1991) 1988 floods in Bangladesh—pattern of illness
    and causes of death. J Diarrhoeal Dis Res 9(4):310–314
Thaipadungpanit J, Wuthiekanun V, Chantratita N, Yimsamran S, Amornchai P, Boonsilp S, Manee-
    boonyang W, Tharnpoophasiam P, Saiprom N, Mahakunkijcharoen Y et al (2013) Short report:
    leptospira species in floodwater during the 2011 floods in the bangkok metropolitan region,
    Thailand. Am J Trop Med Hyg 89(4):794–796
Trevejo RT, Rigau-Perez JG, Ashford DA, McClure EM, Jarquin-Gonzalez C, Amador JJ, de los
    Reyes JO, Gonzalez A, Zaki SR, Shieh WJ et al (1998) Epidemic leptospirosis associated with
    pulmonary hemorrhage—Nicaragua, 1995. J Infect Dis 178(5):1457–1463
Vinetz JM, Glass GE, Flexner CE, Mueller P, Kaslow DC (1996) Sporadic urban leptospirosis. Ann
    Intern Med 125(10):794–798

Chapter 8
Modeling Outbreak Risk Based on the Back Propagation Neural Network (BPNN) Algorithm

8.1 Back Propagation Neural Network (BPNN)

Neural networks, a type of machine-learning algorithm, are efficient mechanisms for inferring relationships and creating models to express the association between input and output parameters (Lee and Hsiung 2009). They are also a significant class of tools for quantitative modeling. The ability to learn relationships between inputs and outputs through data training is a very robust tool characteristic of neural networks (Bai and Jin 2005). From the training process, the neural networks can provide correct prediction values not only for learned examples, but also for any inputs similar to the learned examples. The neural networks method is appropriate for solving large, non-linear, and complex problems of classification and predictive analytics (Srivastava et al. 2013).

Formerly, the concept of neural networks was initiated to develop and test computational analogues of neurons by psychologists and neurobiologists. Simply put, a neural network is a set of connected input/output units in which each connection has a weight associated with it. Throughout the learning phase the network learns by modulating the weights in order to be able to predict the correct class label of the input tuples. Neural network learning is also referred to as connectionist learning due to the connections between units. However, neural networks interlace long training times to adjust the weights and are therefore more appropriate for applications where this is plausible. They also require a number of parameters which are typically best determined empirically.

With several advantages, neural networks have been a very popular machine learning approach for several decades. They have the characteristic of high tolerance of noisy data and the ability to classify patterns on which they have not been trained. They are very flexible to determine the correlation between attributes and classes even in the case when the user may have little knowledge of them. Different from other machine learning algorithms, neural networks are well matched for continuous valued inputs and outputs. They have been proven effective on several real-world or natural data, such as pronouncing English text in the training of a computer, pathology and laboratory medicine, and recognition of handwritten characters. The algorithms of neural networks are spontaneously parallel, which can expedite the computation process. With their popularity, plenty of research has been developed for the extraction of data and rules from trained neural networks. These encouraging developments demonstrate the benefit of neural networks for classification and prediction in data mining.

Several kinds of neural networks and neural network algorithms have been proposed from several research fields. One of the most effective types of neural networks is the BPNN approach, i.e. multi-layer perceptron (MLP) with back propagation, (Han and Kamber 2006; Kanevski et al. 2004; Lee and Hsiung 2009). BPNN has been applied to estimate uncertainties in epidemics and disasters in many studies (Bai and Jin 2005; Cao et al. 2010; Kanevski et al. 2004).

The BPNN is an error back-propagation algorithm for the multi-layer network (Cao et al. 2010). It is provided from the principle that network inputs using several parameters represent an effect of predicting factors, and the outputs will be constructed from a network with the predicting factors. It assesses promising targets by using the network to carry out self-organized learning that ceaselessly projects affected parameters of the nonlinear mapping correlation between the expectations and inputs with a smaller mean square error (MSE). The BPNN component is composed of an input layer, a hidden layer and an output layer that are completely connected together with a layer of neurons between separate layers. Via a number of interconnected neurons, at first weights of the neural network are assigned randomly, and then during the process they are adjusted from a feed-forward to feedback process in order to reduce the MSE. The BPNN steps can be divided into two parts: (1) a forward propagation of information and (2) a backward propagation (BP) of errors that are provided in Sects. 8.2 and 8.3 respectively.

8.2   A Multilayer Feed-Forward Neural Network

The BPNN algorithm conducts learning on a multilayer feed-forward neural network, which iteratively learns a set of weights for prediction of the class label of tuples. A multilayer feed-forward neural network includes an input layer, one or more hidden layers, and an output layer. An example of a multilayer feed-forward network is demonstrated in Fig. 8.1.

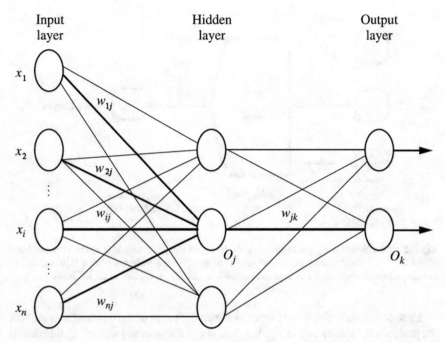

Fig. 8.1 The typical structure of a multilayer feed-forward neural network including the three basic layers (input, hidden, and output layers) (Han and Kamber 2006)

Each layer is constructed from units. The inputs of the BPNN coincide with the attributes measured for each training tuple. The input layer is made up from simultaneously feeding inputs into the units. Subsequently, these inputs pass through the input layer and are weighted and fed to a second layer of "neuron-like" units simultaneously, called as a hidden layer. The outputs of the hidden layer units can be fed into another hidden layer, and so on. Setting the number of hidden layers is capricious, but it is usually defined by the user. The weighted outputs of the last hidden layer are input to units forming the output layer, which evolves the network's forecasting for given tuples.

The units in the input layer are called input units, while the units in the hidden layers and output layer are referred to either as neurodes, due to their symbolic biological basis, or as output units. As in Fig. 8.1 the multilayer neural network may have two layers of output units, which usually is called a two-layer neural network (the input layer is not counted because it serves only to pass the input values to the next layer). Similarly, a network composed of two hidden layers is called a three-layer neural network, and so on. The network is feed-forward in that none of the weights cycle back to an input unit or to an output unit of a previous layer. It is completely linked in that each unit feeds input to each unit in the next forward layer.

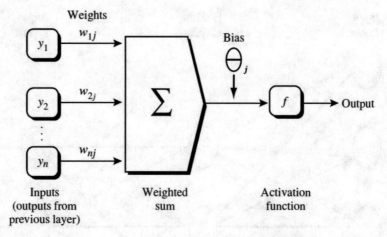

Fig. 8.2 For a hidden or output layer unit j: the inputs to unit j are outputs from the previous layer which are multiplied by their consistent weights so as to build a weighted sum, which is added to the bias associated with unit j. A nonlinear activation function is applied to the net input

Each output unit also acts as input as a weighted sum of the outputs from units in the previous layer as shown in Fig. 8.2. It employs a nonlinear (activation) function to the weighted input. Multilayer feed-forward neural networks can estimate the class prediction with a combination of nonlinear functions into the inputs. Viewing from a statistical point, they conduct nonlinear regression. Given enough hidden units and enough training samples, multilayer feed-forward networks are able to estimate any function very accurately.

To begin the training process, firstly the network topology should be considered by initially defining the number of units in the input layer, the number of hidden layers (if more than one), the number of units in each hidden layer, and the number of units in the output layer.

Before feeding the input tuples, the normalization of the input values for each attribute observed in the training tuples can assist to speed up the learning process. The range of input values is commonly normalized between 0 and 1. Discrete-valued attributes may be encoded such that there is one input unit per domain value. For example, 'A' may be assigned to three input units if an attribute A includes three possible or known values, namely $[a_0, a_1, a_2]$. Similarly, input units may be defined as I_0, I_1, I_2. Each unit is initialized to zero. If $A = a_0$, then I_0 is set to 1. If $A = a_1$, I_1 is set to 1, and so on.

Neural networks can be applied for both classification (to forecast the class label from a given tuple) and prediction (to estimate coherent-valued output). In classification, one output unit may be employed to indicate two classes (where value 1 represents one class, and value 0 represents the other), and if there are more than two classes, then we can use one output unit per class.

The best number of hidden layer units is not explicit and there are no clear rules. Therefore, a trial-and-error process is commonly applied for the network design, which may affect the accuracy of the resulting trained network. The initial weights defining the network may also affect the accuracy of results. If the resulting accuracy is not considered acceptable from training a network once, the training process will commonly repeat with a different set of initial weights or a different network topology.

To decide an acceptable network, a cross-validation method usually is applied for accuracy estimation. A number of automated techniques have been provided which search for a "good" network topology, and these typically apply a hill-climbing method that starts with an initial structure that is selectively modified.

8.3   Back Propagation

Back propagation learns by repeatedly processing a data set of training tuples. It compares the prediction value of the network for each tuple with the actual value from a known target. The target value may be a continuous value (for prediction) or the known class label of the training tuple (for classification problems). The weights of each training tuple are adjusted in order to minimize the mean squared error between the network's prediction value and the actual target value. These modifications are called "back propagation", because they are conducted in a "backwards" direction from the output layer via each hidden layer down to the first hidden layer. Although it is not guaranteed, ordinarily the learning process will stop by the time the weights converge in the end. The components of back propagation proposed in the form of inputs, outputs, and errors may appear awkward at first sight for someone who first begins to study the neural network. However, after becoming familiar with the neural network, one will realize that each step is inherently simple

The steps of BPNN can be simply explained as below.

Input:

(1) D is a data set containing the training tuples as well as their related target values;
(2) l is the learning rate;
(3)   *network* is a multilayer feed-forward network.

Output: a trained neural network.

Approach:

(1) Define all initial weights and biases in the network;

(2) while preparing inputs, the terminating condition is not satisfied{

(3)   for each training tuple X in D {

(4)   // Propagate the inputs forward:

(5)   for each input layer unit j{

(6) $O_j = I_j$; // output of an input unit is its actual input value

(7)   for each hidden or output layer unit j{

(8) $I_j = \sum_i w_{ij} O_i + \theta_j$; //calculate the net input of unit j with respect to the previous layer i

(9) $O_j = \dfrac{1}{1 + e^{-I_j}}$; }}} //calculate the output of each unit j

(10)   // Backpropagate the errors:

(11)   foreach unit j in the output layer

(12) $Err_j = O_j(1 - O_j)(T_j - O_j)$; // calculate the error

(13)   foreach unit j in the hidden layers, from the last to the first hidden layer

(14) $Err_j = O_j(1 - O_j)\sum_k Err_k w_{jk}$; // calculate the error with respect to the next highest layer k

(15)   foreach weight w_{ij} in $network$ {

(16) $\Delta w_{ij} = (l)Err_j O_i$; // weight adjustment

(17) $w_{ij} = w_{ij} + \Delta w_{ij}$; // weight improvement

(18)   foreach bias θ_j in $network$ {

(19) $\Delta \theta_j = (l)Err_j$; // bias adjustment

(20) $\theta_j = \theta_j + \Delta \theta_j$; / bias improvement

(21)   }}}

Determining the initial weights, the weights of the network are initialized with small random numbers, for example, ranging of -1.0 to 1.0 or -0.5 to 0.5. Each unit has a bias associated with it as expressed in Fig. 8.2. This bias is also initialized to small a random number.

While X is each training tuple, the process of each tuple can be described by the following steps.

In a multilayer feed-forward network, in the beginning the training tuple is fed into the input layer of the network to propagate the inputs forward. The inputs pass through the input units without changing. That is, for an input unit, j, its output, O_j, is equal to its input value, I_j. Subsequently, the net input and output of each unit in the hidden and output layers are computed. A linear combination of the inputs is used to compute the net input to a unit in the hidden or output layers. Figure 8.2 demonstrates a hidden layer or output layer unit as well as its weight and bias. Each such unit has a number of inputs which are in fact the outputs of the units connected to it in the previous layer. A weight is linked to each connection. The sum of each unit-connected weight is multiplied with each input for computing the net input to the unit. Defined j is a unit in a hidden or output layer, the net input, I_j, to unit j is

$$I_j = \sum_i w_{ij} O_i + \theta_j \tag{8.1}$$

where w_{ij} is defined as the weight of the connection between unit i in the previous layer and unit j; O_i is the output of unit I of the previous layer; and θ_j is the unit bias. The bias executes as a threshold in that it acts to modify the activity of the unit. As per the above explanation, $y_1, y_2, \ldots, y_n$ represent the inputs to unit j. When the first hidden layer contains unit j, the input tuple $(x_1, x_2, \ldots, x_n)$ would coincide with these inputs.

Each unit from the output and hidden layers obtains its net input and then applies an activation function to it, as illustrated in Fig. 8.2. The sigmoid, or logistic function, is employed to represent the motivation of the neuron. Given the net input I_j to unit j, subsequently the output O_j of unit j, is provided as

$$O_j = \frac{1}{1 + e^{-I_j}} \tag{8.2}$$

Also, the function is referred to as a *squashing function*, since it applies a large input domain onto the smaller range of 0–1. Because the logistic function is differentiable and nonlinear, the back propagation algorithm can model the classification problems, which are linearly indivisible.

The output values, O_j, of each hidden layer are computed up to and including the output layer in order to find the prediction of the network. Practically applied, restoring or saving the moderate output values of each unit is recommended as they are required again later when back propagating the error, which will lead to reducing the amount of computation required substantially.

The error is propagated backward by modifying the weights and biases to reduce the error of the network's prediction. For a unit j in the output layer, the error Err_j is represented by

$$Err_j = O_j(1 - O_j)(T_j - O_j) \tag{8.3}$$

where O_j is the actual output from unit j, and T_j is the reference target value of the defined training tuple. Note that $O_j(1 - O_j)$ is the derivative of the logistic function.

We consider the weighted sum of the errors of the units connected to unit j in the next layer for computing the error of a hidden layer unit j. Thus, the error of a hidden layer unit j is

$$Err_j = O_j(1 - O_j) \sum_k Err_k w_{jk} \tag{8.4}$$

where w_{jk} is the weight of the connection between unit j and a unit k in the next layer, and Err_k is the error of unit k.

Updating the weights and biases is able to reflect the propagated errors. A weight is modified by using the following equations. Given Δw_{ij} is the change in weight w_{ij}:

$$\Delta w_{ij} = (l) Err_j O_i \tag{8.5}$$

$$w_{ij} = w_{ij} + \Delta w_{ij} \tag{8.6}$$

Given that the learning rate is the variable l, the range of l is commonly a value between 0.0 and 1.0. A target of back propagation is to minimize the mean squared distance between the network's class prediction and the known target value of the tuples by using a method of gradient descent to search for a set of weights that fits the training data. The learning rate, a variable of back propagation, assists in avoiding being stuck at a local minimum decision space (i.e., where the weights begin to converge, but are not the optimal solution) and supports finding the global minimum. The learning will happen at a very slow pace when the learning rate is a too small value. In contrast, if the learning rate is too large, then oscillation from in sufficient solutions may occur. The learning rate is assigned by a rule of thumb. The learning rate is given at $1/t$, where t is the number of iterations through the training set so far.

We employ the following equations below to update biases, where $\Delta \theta_j$ is the change in bias θ_j:

$$\Delta \theta_j = (l) Err_j \tag{8.7}$$

$$\theta_j = \theta_j + \Delta \theta_j \tag{8.8}$$

As per the above explanation, the weights and biases are updated after the presentation of each tuple, and this refers to case of having updated the value. On the other hand, the increments of the weight and bias could be enhanced in variables, and after all of the tuples in the training set have been provided the weights and biases are adjusted.

This latter procedure is called epoch updating, where an epoch refers to an iteration of the training set. Theoretically, the mathematical derivation of back propagation employs epoch updating. However, practically, case updating is more ordinary since it tends to yield better and more accurate results.

The training process will stop when:

(1) All Δw_{ij} and $\Delta \theta_j$ of the previous epoch were too small and less than some defined threshold, or

(2) The percentage of tuples misclassified in the previous epoch is below some threshold, or

(3) A number of epochs meet a pre-defined value.

Practically speaking, several hundreds of thousands of epochs are usually required before the weights will converge to a solution.

The computational effectiveness relies on the time spent training the network. Given |D| tuples and w weights, each epoch requires $O(|D| \times w)$ time. However, the worst-case scenario may express the number of epochs, which are exponential lin n, the number of inputs. Practically, the time the network uses to converge varies, and we can employ some techniques to assist speeding up the training time. For example, there is a technique known as simulated annealing, which can ensure convergence to a global optimum.

8.4   Modeling Outbreak Risk Based on BPNN

8.4.1   Initial Analysis of Input and Reference Data

According to the outbreak risk ratio, we found that during flooding the outbreak of diarrhea is far more severe than that of conjunctivitis and leptospirosis. Therefore, this study attempted to model the outbreak risk of diarrhea by using flood-related parameters.

Analyzing the model input parameters, Table 8.1 shows the RR of each district during weeks 36–52 derived from the diarrheal morbidity as reported by hospitals through the means of three main parameters: population density, flood duration, and DO. Theoretically, an area having high population density, long flood duration, or lower DO should have a high risk of disease and vice versa. The mean values of each district in Table 8.1 suggest that all three parameters affect the risk of diarrheal outbreak. For example, although the Bang Pa-in and Phra Nakhon Si Ayudhya districts had the worst DO quality and the highest population density respectively, their outbreak risks were moderate due to the comparatively short duration of their floods. Even though the Bang Sai district had the longest flood duration, its outbreak risk was also moderate due to having the lowest population density.

To prepare the reference output data for the network training, we distributed the resulting outbreak risks (Table 8.1) from the location of district hospitals in the study

Table 8.1 The risk ratio (RR) of diarrheal outbreak due to flooding and the related parameters for the eight districts in the study area

District	Population density (people/km^2)	Mean flood duration (days)	Mean DO (mg/L)	Diarrheal morbidity	Risk ratio
Sena	307.14	67	3.6	716	1.48
Bang Ban	251.40	53	3.9	183	1.45
Uthai	272.77	40	3.3	462	1.32
Bang Sai	119.51	75	4.3	138	1.23
Bang Pa-in	380.38	35	3.0	711	1.15
Phra Nakhon Si Ayudhya	1180.86	30	3.6	1651	1.14
Bang Chai	188.31	44	3.9	292	0.99
Bang Pahan	315.46	44	3.7	195	0.82

area and its surroundings by using IDW interpolation. The RR of each district in Table 8.1 was spatially spread over the study area based on IDW to produce the target values for BPNN training and testing. We defined the RR position as the location of the hospital in each district, because in general diarrheal patients should go to the district hospital nearest to their residences. Thus, it is plausible to employ IDW to spatially interpolate the outbreak risk by considering the distance from the training/testing point to the district hospital. In addition, we also added a few RRs of district hospitals outside the study area for better interpolation near the boundaries of the study area. The reference risk map and the locations of the district hospitals are shown in Fig. 8.3.

8.4.2   Results of Training and Testing the BPNN

In this study we used measures of flood duration, DO, and population density at various locations as the input dimensions, and they were normalized to lie in a fixed range, from zero to one, by subtracting the minimum value and dividing by the range between the maximum and the minimum values (Witten and Frank 2005). With the spatial resolution of our input parameters, which were the 50-m Radarsat-2 data, each tuple fed into the network was defined as each center point of a 50-m grid (point-based) (Yomwan et al. 2013), and each value for each input dimension was determined by the pixel value of the flood duration, DO, and population-density map in which its tuple fell. We used the RR of disease outbreak based on surveillance reports as reference data for training. After the model was trained, we used it to predict RR for a new set of tuples.

The RR of each district was spatially spread over the study area based on IDW to produce the target values for BPNN training and testing. We defined the RR position

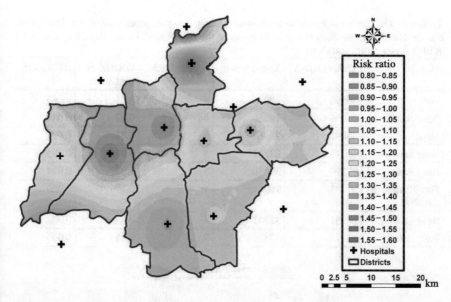

Fig. 8.3 The spatial distribution of the risk ratio (RR) of diarrheal outbreak due to the 2011 major flood in the study area (an RR less than or equal to 1 indicates no outbreak or no risk; otherwise, it indicates the possibility of an outbreak, with outbreak intensity increasing with larger values of RR)

as the location of the hospital in each district, because in general diarrheal patients should go to the district hospital nearest to their residences. Thus, it is plausible to employ IDW to interpolate the spatial distribution of outbreak risk by considering the distance from the training/testing point to the district hospital. In addition, we also added a few RRs of district hospitals outside the study area for better interpolation near the boundaries of the study area.

The map resolution was defined by the best spatial resolution of our input parameters, which was the 50 m × 50 m resolution. We had a total of 569,122 points (pixels). These points were divided into two groups: 66% for training and 34% for testing, which is the default setting for splitting learning data in WEKA (Bouckaert et al. 2012), a popular open-source machine-learning software developed by the University of Waikato, New Zealand.

During the training process, setting a suitable number of neurons in the hidden layer is very important. If the number of hidden neurons exceeds an appropriate value, the computation is time-consuming and unstable. On the other hand, if the number of neurons is too small, the training process may not converge (Cao et al. 2010). Due to the large number of data tuples, we decided to split the input data by district. The numbers of hidden-layer neurons for each district, which ranged from 3 to 9, are shown in Table 8.2.

We examined the precision of the BPNN model by plotting a graph between the predicted risk and reference risk. Figure 8.4 represents some scatter plots of Bangpahan, Bang Sai and Bangban with R values of 0.89, 0.91 and 0.83 respectively.

Table 8.2 The number of inputs for training and testing, the appropriate number of hidden layers, and the correlation coefficients and RMS errors of the BPNN model prediction computed using WEKA (open-source software)

District	Trained inputs	Tested inputs	Hidden layers	Correlation coefficient	RMS error
Sena	56998	29096	3	0.81	0.065
Bang Ban	36036	18662	8	0.83	0.038
Uthai	44891	23362	8	0.81	0.037
Bang Sai	43371	22501	5	0.93	0.035
Bang Pa-in	62770	32086	9	0.83	0.014
Phra Nakhon Si Ayudhya	31105	16014	7	0.79	0.029
Bang Chai	66051	33953	4	0.83	0.048
Bang Pahan	34398	17828	7	0.88	0.060

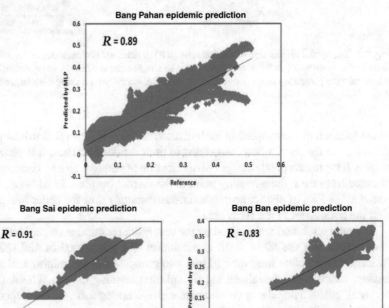

Fig. 8.4 The scatter plot between the diarrhea epidemic risk derived from hospital morbidity report (reference) and BPNN prediction risk in Bang Pahan, Bang Sai and Bang Ban district

To evaluate the prediction success, we assessed the accuracy by using correlation coefficients (R-values) and RMS errors. RMS error is the most commonly used measure in mathematical techniques to assess how close the predicted values from the model are to the actual or reference values, while the correlation coefficient can measure the statistical relationship between the test instances and the actual or reference values and indicate how well the data points or instances fit the linear line (Witten and Frank 2005). The results of our BPNN predictions in Table 8.1 show that the correlation coefficients range from 0.79 to 0.93 (the correlation coefficient ranges from 1, for perfectly correlated results, to 0, when there is no correlation, to −1 when the results are perfectly correlated negatively), the RMS errors range from 0.014 to 0.065, and the predicted RR range from 0.82 to 1.57. The high correlation coefficients indicate that our results have a very good correlation with the reference or actual data, and the low RMS errors indicates that the predicted RR derived from the model and the reference RR derived from the morbidity data are close in value. An RR resolution (the difference of two RR values) below 0.1 indicates an indifferent risk for classifying the strength of an epidemiological association (Craun et al. 2006). This can imply that the range of our RMS errors is less than the adequate RR resolution to classify the level of outbreak risk. As discussed above, we can conclude that our predictive model for each district can give accurate predictions.

8.5 Summary

The back propagation neural network (BPNN) approach, i.e. multi-layer perceptron (MLP) with back propagation, is considered one of the most effective types of neural networks. The BPNN algorithm performs learning on a multilayer feed-forward neural network, which includes the three basic layers (input, hidden, and output layers). Analysis of the model input parameters reveals that flood duration, dissolved oxygen, and population density all affect the risk of diarrheal outbreak. In this study we used measures of flood duration, DO, and population density at various locations as the input dimensions, which were divided into two groups: 66% for training and 34% for testing to be processed in WEKA (a popular open-source machine-learning software). The results of our BPNN predictions show that the correlation coefficients range from 0.79 to 0.93 (the correlation coefficient ranges from 1, for perfectly correlated results, to 0, where there is no correlation, to −1 when the results are perfectly correlated negatively), RMS errors range from 0.014 to 0.065, and the predicted RR range from 0.82 to 1.57. The high correlation coefficients indicate that our results have a very close correlation with the reference or actual data, and the low RMS errors indicate that the predicted RR derived from the model and the reference RR derived from the morbidity data are close to each other. We therefore can conclude that our predictive model for each district can give accurate predictions.

References

Bai YP, Jin Z (2005) Prediction of SARS epidemic by BP neural networks with online prediction strategy. Chaos, Solitons Fractals 26(2):559–569

Bouckaert RR, Frank E, Hall M, Kirkby R, Reutemann P, Seewald A, Scuse D (2012) WEKA Manual for Version 3.7.6 [Internet]. University of Waikato, Hamilton, New Zealand

Cao CX, Chang CY, Xu M, Zhao JA, Gao MX, Zhang H, Guo JP, Guo JH, Dong L, He QS et al (2010) Epidemic risk analysis after the Wenchuan Earthquake using remote sensing. Int J Remote Sens 31(13):3631–3642

Craun GF, Calderon RL, Wade TJ (2006) Assessing waterborne risks: an introduction. J Water Health 4(Suppl 2):3–18

Han J, Kamber M (2006) Data mining: concepts and techniques. Elsevier, San Francisco

Kanevski M, Parkin R, Pozdnukhov A, Timonin V, Maignan M, Demyanov V, Canu S (2004) Environmental data mining and modeling based on machine learning algorithms and geostatistics. Environ Model Softw 19(9):845–855

Lee CJ, Hsiung TK (2009) Sensitivity analysis on a multilayer perceptron model for recognizing liquefaction cases. Comput Geotech 36(7):1157–1163

Srivastava PK, Han DW, Ramirez MR, Islam T (2013) Machine learning techniques for downscaling smos satellite soil moisture using MODIS land surface temperature for hydrological application. Water Resour Manage 27(8):3127–3144

Witten IH, Frank E (2005) Data mining: practical machine learning tools and techniques. Elsevier, San Francisco

Yomwan P, Cao C, Rakwatin P, Suphamitmongkol W, Tian R, Saokarn A (2013) A study of waterborne diseases during flooding using Radarsat-2 imagery and a back propagation neural network algorithm. Geomatic, Nat Hazards Risk:1–19

Chapter 9
Application of Surveillance of Communicable Disease Risk Using Expert Systems

In this chapter an example is illustrated which applies the knowledge of flood identification methods already described in Part I and the knowledge of waterborne diseases for developing a system that can provide basic predictions for patients. The Expert System (ES) was implemented as an example, because it has no complexity when being employed, thus resulting in less time-consuming calculation.

9.1 Introduction to the Expert System

Dating back many decades, expert systems (ES) were founded in the development of artificial intelligence (AI). The assertion that expertise, defined as knowledge related to specific activities, can be passed from humans to computers is the basis of the concept developed by the AI community. This expertise is accessible to humans whenever they require the information and can be stored. In applicable situations, the computer would be capable of evaluating conditions and formulating suitable responses to presented questions. The computer's reply may take the form of a rationalization or directive, with the prospect of providing additional insights into the rational processes behind the reply (Thomson 2004). An ES is a computer application capable of providing proficient answers to complicated problems with immense flexibility (Buchanan and Duda 1982). At the same time, the explanations provided are innately evident and rooted in heuristics. Additionally, adjustment and enhancement of solutions derived from one situational case to address various problems in the future are possible by applying case-based reasoning strategies (CBR) in ES. Utilizing CBR makes it feasible to assemble the understanding of human specialists in a variety of cases and file these in a database to function as a knowledge base for managing enquiries involving comparable situations in the future. Consequently, the ES explores its database for related cases when a problem is given. The solutions provided in such a case can be utilized for the current problem if a match is discovered. The new case could be added to the database provided the solutions were

© Higher Education Press and Springer Nature Singapore Pte Ltd. 2021   135
C. Cao et al., *Environmental Remote Sensing in Flooding Areas*,
https://doi.org/10.1007/978-981-15-8202-8_9

accurate and successful, which would expand the knowledge base further. Even information for cases where the solutions proposed were unsuccessful could be filed for future reference as an indicator for future cases employing the knowledge database (Aamodt and Plaza 1994).

The progression of computer software began in the early 1950s with a concentration on numerical systems. In the early part of the 1970s, people started to rely increasingly on computers for handling data processes. MYCIN was a rules-based system created by Stanford University in the 1970s. It remains representative of the state of the art Expert System, despite the fact that it was developed many years ago. MYCIN was intended to aid consistent diagnosis of the probable cause of patient infection and offer suggestions for corrective treatment. Its rule base was capable of classifying various meningitis infections and contained 450 rules. The system was successful and confirmed that, with a relatively straight forward representation of rules in the form of if-then-else equations, somewhat complex but decisive domain areas, such as meningitis infection diagnosis, were possible. Researchers began exploring the possibility of artificial intelligence in the 1980s. Many Expert System tools started to materialize around this time. The United States, Japan, and numerous other nations in Europe started investing in the field of artificial intelligence and Expert Systems. The widening application of computers came to include numerical calculations, as well as the ability to possess knowledge and subsequently improve the operation of computers. This comprises the primary function of artificial intelligence. Natural language processing, symbol processing, rule-based systems, and logic systems are elements included in the principle research field of artificial intelligence.

It has also been suggested as a practical application to be employed in the medical field (Abidi and Manickam 2002). Electronic medical records (EMR), produced out of custom, could provide an exceptional source of data for CBR analysis. Also proposed for the research is the automated formatting of XML-based EMR data for CBR applications, which will result in a diagnostic system for medical practices. A noteworthy element of the process entails the use of rule-based reasoning (RBR) to form precise analyses employing the official knowledge base. Similarly, there exists research (Montani and Bellazzi 2002) supporting the utilization of multi-modal reasoning (MMR) in an effort to assimilate a more extensive range of accessible kinds of knowledge. This would permit utilization of a mixture of RBR and CBR, affording more exact deductions to be made in terms of context, which would result in improved recovery of information and decision-making. In tests concerning the supervision of patients with diabetes, this approach has been revealed to work efficiently.

The capability to utilize data from a human expert and convert that data into a set of rules employing constructs such as If-Then comprises the principle function of rule-based expert systems. Subsequently, the system is capable of applying its rules to existing cases to provide appropriate replies. An approach for examining information and formulating judgments through a combination of the rule base and knowledge base is created by the computer program's capacity for producing inferences (Shu-Hsien 2005). The advancement of efficient Expert Systems is constantly focused on the regulation of a knowledge base. The standard integration of Expert System components is provided in Fig. 9.1 (Chen et al. 2012). Systems in which computers attempt to comprehend and utilize human knowledge are classified as knowledge-based systems (KBS). KBS usually comprise four primary elements and are found in artificial intelligence (Wiig 1994). They are a knowledge base, a user interface, a tool allowing knowledge engineering, and an inference engine. Then again, any system consisting of all information technology applications within an organization capable of managing the organization's knowledge base can be classified as 'KBS' as well. Therefore, the term applies to ES and other rule-based systems, group-ware and database management systems (Laudon and Laudon 2002). As knowledge evolves, the knowledge engineers are able to create, modify, or delete the contents of the knowledge base. Recompiling and redeploying the application is unnecessary while maintaining the ability to update knowledge content. The problem-solving rules assembled and converted by human experts in a specified problem domain are contained in the knowledge base. The interpretation of knowledge in a knowledge base is ordinarily presented in the following form:

IF <conditions>

THEN <action list_1>

ELSE <action list_2>

Consequently, either action list 1 or action list 2 is applied depending on the circumstances. A rule-based ES is characterized as including data acquired from a human authority and the representation of that same data in the form of rules, such as IF–THEN (Shu-Hsien 2005). Subsequently, the rule can be applied to data processes in order to arrive at suitable deductions.

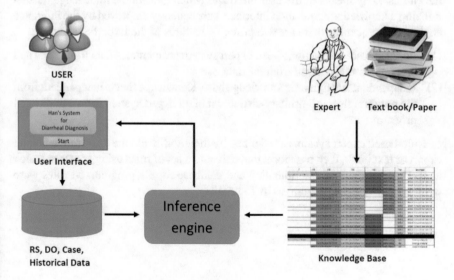

Fig. 9.1 The standard structure of an Expert System for Diagnosis

9.2   Integrated Expert System for Surveillance of Communicable Disease Risk

9.2.1   Expert System for Diagnosis Design

The most important process of the expert system is the knowledge acquisition. The process of acquiring, organizing and studying knowledge is classified as knowledge acquisition. In the study, the initial step prior to development of the system required an analysis of prevailing guidelines for the diagnosis of diarrhea infection. For example, a particular concern is water polluted with human feces originating from public sewage, septic tanks, and latrines. The microorganisms that can cause diarrhea are also present in animal feces. Diarrhea is transmittable from person to person and exasperated by inadequate hygienic practices. When prepared or stored in unhygienic conditions, food is another significant source of diarrhea. Fish and other seafood collected from contaminated water sources may contribute to the spread of diarrhea, as well as water contamination of fresh food in the course of irrigation. Bowel movements occurring more than three times in a 24-h period, possibly with vomiting or dehydration, was set as the initial criteria for patients. Clinical data was grouped into several categories, including (1) patient history records together with the principal complaint of patients and associated risk factor(s) for diarrhea, (2) Physical assessment, and (3) diagnostic tests for the presence of diarrhea. The knowledge base, inference engine, user interface and storage data comprise the 4 components of the system (historical data being RS, DO, cases, etc.). All units work together systematically as in Fig. 9.1. An expert system involves initial communication with the end users by means of the user interface, which gathers the knowledge needed utilizing a knowledge acquisition interface, subsequently employed by the inference engine. Consequently, there are two principle features of the Expert System:

(1) Frequent user interactions—An expert system accumulates data from users and prepares results derived from the data.
(2) Independent and dynamic knowledge base-Knowledge that is independent from the program flow is a primary distinction of an expert system from a traditional program.

Rule-based expert systems are defined by their ability to take data from a human expert or text book, then transform that data into a set of rules using constructs such as If-Then for the rule determination and database for the program. 54 rules were written for use in the program as in Table 9.1.

Table 9.1 The 54 rules of the expert system for diagnosis

RULE	IF	AND	AND	AND	AND		Then	Recommend
1	Watery = 1 or 2 in past 24 h	V	DO ≥ 6	HF	HD	WF	No risk	1
2			DO ≥ 6	HF	HD	WF	No risk	1
3		V	4 ≤ DO < 6	HF	HD	WF	No risk	1
4			4 ≤ DO < 6	HF	HD	WF	No risk	1
5		V	2 ≤ DO < 4	HF	HD	WF	No risk	1
6			2 ≤ DO < 4	HF	HD	WF	No risk	1
7		V	DO ≥ 6	HF		WF	No risk	1
8			DO ≥ 6	HF		WF	No risk	1
9		V	4 ≤ DO < 6	HF		WF	No risk	1
10			4 ≤ DO < 6	HF		WF	No risk	1
11		V	2 ≤ DO < 4	HF		WF	No risk	1
12			2 ≤ DO < 4	HF		WF	No risk	1
13		V	DO ≥ 6	HF	HD		No risk	1
14			DO ≥ 6	HF	HD		No risk	1
15		V	4 ≤ DO < 6	HF	HD		No risk	1
16			4 ≤ DO < 6	HF	HD		No risk	1
17		V	2 ≤ DO < 4	HF	HD		No risk	1
18			2 ≤ DO < 4	HF	HD		No risk	1
19	Watery = 3 in past 24 h	V	DO ≥ 6	HF	HD	WF	Risk	2
20			DO ≥ 6	HF	HD	WF	Risk	2
21		V	4 ≤ DO < 6	HF	HD	WF	Risk	2
22			4 ≤ DO < 6	HF	HD	WF	Risk	2
23		V	2 ≤ DO < 4	HF	HD	WF	Risk	2
24			2 ≤ DO < 4	HF	HD	WF	Risk	2
25		V	DO ≥ 6	HF		WF	Risk	2
26			DO ≥ 6	HF		WF	Risk	2
27		V	4 ≤ DO < 6	HF		WF	Risk	2
28			4 ≤ DO < 6	HF		WF	Risk	2
29		V	2 ≤ DO < 4	HF		WF	Risk	2
30			2 ≤ DO < 4	HF		WF	Risk	2
31		V	DO ≥ 6	HF	HD		Risk	2
32			DO ≥ 6	HF	HD		Risk	2
33		V	4 ≤ DO < 6	HF	HD		Risk	2
34			4 ≤ DO < 6	HF	HD		Risk	2
35		V	2 ≤ DO < 4	HF	HD		Risk	2

(continued)

Table 9.1 (continued)

RULE	IF	AND	AND	AND	AND		Then	Recommend
36			$2 \leq DO < 4$	HF	HD		Risk	2
37	Watery > 3 in	V	$DO \geq 6$	HF	HD	WF	Risk	2
38	past 24 h		$DO \geq 6$	HF	HD	WF	Risk	2
39		V	$4 \leq DO < 6$	HF	HD	WF	Risk	2
40			$4 \leq DO < 6$	HF	HD	WF	Risk	2
41		V	$2 \leq DO < 4$	HF	HD	WF	Risk	2
42			$2 \leq DO < 4$	HF	HD	WF	Risk	2
43		V	$DO \geq 6$	HF		WF	Risk	2
44			$DO \geq 6$	HF		WF	Risk	2
45		V	$4 \leq DO < 6$	HF		WF	Risk	2
46			$4 \leq DO < 6$	HF		WF	Risk	2
47		V	$2 \leq DO < 4$	HF		WF	Risk	2
48			$2 \leq DO < 4$	HF		WF	Risk	2
49		V	$DO \geq 6$	HF	HD		Risk	2
50			$DO \geq 6$	HF	HD		Risk	2
51		V	$4 \leq DO < 6$	HF	HD		Risk	2
52			$4 \leq DO < 6$	HF	HD		Risk	2
53		V	$2 \leq DO < 4$	HF	HD		Risk	2
54			$2 \leq DO < 4$	HF	HD		Risk	2

Notes V, vomiting; DO, dissolved oxygen; HF, lives in the flooded area; HD, have dry skin or dry lips; WF, works in a factory; 1, go home and rest; 2, go to the hospital

9.2.2 Expert System for Diagnosis Evaluation

An experiment was conducted by collecting the history of 100 current patients and the patients in 2011. Then, the results of the diagnoses and advice were obtained from the expert system for medical examination (as Figs. 9.2, 9.3, 9.4).

Then an assessment of expert system for diagnosis was made by measuring precision and recall.

The assessment showed that the system could provide diagnosis and advice to diarrhea patients by using a database from remote sensing with a precision of 97% and a recall of 95% compared with the answers of professional physicians (Fig. 9.5). The assessment analysis of the Expert System showed that each rule had the precision of 100%. However, when applying this with the 54 rules, some rules resulted in a similar diagnosis, but the level of symptom severity was not equal, resulting in a demonstration of the same diagnosis but in regards to unequal severity. The experts considered this to be incorrect and that it affected the accuracy of the system. The error of some recall depended on the result of the precision error, and some were

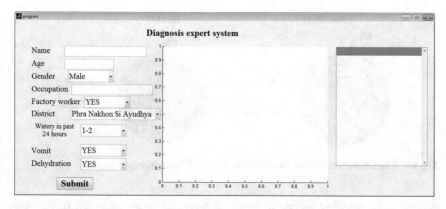

Fig. 9.2 Knowledge base editor of the expert system for diarrheal diagnosis

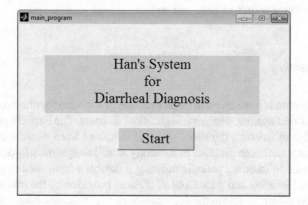

Fig. 9.3 User interface of the expert system for diarrheal diagnosis

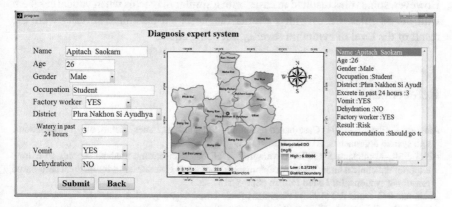

Fig. 9.4 The output of the expert system for diagnosis showing the results of a risk analysis and a map of the correlation between DO level and the user

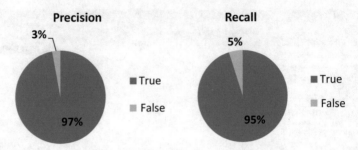

Fig. 9.5 The assessment showed the precision and recall from the expert system

derived from defective advice in terms of food and water when the patient was dehydrated. The experts considered that each patient should be given correct advice regarding proper amounts of food and water.

9.3   Summary

The expert system is introduced to be an example of a system for diagnosis of people who live in a flooding area, which has high risk of sickness. The Expert System applies the knowledge of flooding duration, which is extracted from satellite data, and the knowledge of waterborne diseases in the study area. The system afforded diagnostic data and advice to diarrhea patients utilizing a database from remote sensing with an accuracy of 96.89% and a recall of 95.39% as provided by the assessment. This information was comparable to the determinations of physicians. Each rule possessed a precision of 100%, as provided by an appraisal examination of the expert system. However, some rules resulted in cases with a similar diagnosis when validating them with the 54 rules. Cases with a similar diagnosis but dissimilar severity were the result of the level of symptom severity not being equal.

References

Aamodt A, Plaza E (1994) Case-based reasoning: foundational issues, methodological variations, and system approaches. AI Commun 7(1):39–59

Abidi SSR, Manickam S (2002) Leveraging XML-based electronic medical records to extract experiential clinical knowledge. An automated approach to generate cases for medical case-based reasoning systems. Int J Med Inform 68(1–3):187–203

Buchanan BG, Duda RO (1982) Principles of rule-based expert systems. Stanford University

Chen Y, Hsu C-Y, Liu L, Yang S (2012) Constructing a nutrition diagnosis expert system. Expert Syst Appl 39(2):2132–2156

Laudon KC, Laudon JP (2002) Essential of management information system, 5th edn. Prentice Hall, Englewood cliffs, NJ

Montani S, Bellazzi R (2002) Supporting decisions in medical applications: the knowledge management perspective. Int J Med Inform 68(1–3):79–90

Shu-Hsien L (2005) Expert system methodologies and applications—a decade review from 1995 to 2004. Expert Syst Appl 28(1):93–103

Thomson AJW (2004) A web-based expert system for advising on herbicide use in Great Britain. Application note. Comput Electron Agric 42:43–49

Wiig KM (1994) Knowledge management: the central management focus for intelligent-acting organizations. Schema Press

Chapter 10
Conclusions and Discussions

10.1 Summary of Major Results

10.1.1 Flood Identification

The first study for flood water identification is the extraction of water by vegetation index and histogram thresholding methods, which are popular methods for detecting flood areas based on multispectral images and SAR images. The second study implement the SVM and PF. This study presents an SVM-based PF approach. The method employed a PF to estimate three key parameters of the SVM training model (the kernel function, the insensitive loss function, and the regularization parameter) to search for an appropriate value of these parameters before passing them to an SVM training model. We took advantage of the recursive technique of the PF and the updated particle to find a suitable parameter value. The modified parameter values have characteristics that are more appropriatefor training samples than those of the original model, which reflects the strong correlation between these parameters and the dataset. The SVM-PF is demonstrated to be better in regards to accuracy, precision, and recall analysis compared to the original methods. Although the particle filter requires more complex computations than k-fold cross-validation, the SVM-PF can be applied to several applications, such as flood monitoring and flood loss estimation, which require precise knowledge of flood areas. Therefore, the proposed method was applied to two kinds of application. The first application is water identification in flooding areas using SAR imagery and data acquired on December 4, 2011 in Ayutthaya province, Thailand. The main results of the study on flood identification in this book can be summed up in the following points:

(1) The flooded map series derived from a combination method of water body delineation of passive and active remote sensing imagery illustrate that the 2011 Thailand major flood in our study area began approximately in the beginning of September 2011 and continuously increased in October until its peak in the

© Higher Education Press and Springer Nature Singapore Pte Ltd. 2021 145
C. Cao et al., *Environmental Remote Sensing in Flooding Areas*,
https://doi.org/10.1007/978-981-15-8202-8_10

middle of November. Subsequently, the flood gradually abated from the beginning to the end of December 2011. The validation of classified flooded results expresses that the integration technique of NDWI, thresholding and visualization is appropriate for ongoing flood monitoring based on multi-satellite imagery.

(2) The flood duration map estimated from the flooded map series indicates that the inundated areas covered the entire study area at its peak. The longest inundated period was over four months, while the shortest was barely one week. The relatively long periods of inundation areas tend to be agricultural areas mainly located in low areas.

(3) A program was developed for the SVM classifier using a Gaussian radial basis function kernel for two classes (using in water identification application) and multi-classes (using in wetland class separation).

(4) The relationship between the dataset and the initial SVM training parameters was obtained, which in this experiment presents the average weights of every particle in each iteration time.

(5) We developed a program for combining the process of the SVM and the PF algorithms, which became the proposed method (the SVM-PF method).

(6) We implemented the proposed method in water identification of flooding areas in which SAR imagery was applied and only two classes considered: water and non-water.

10.1.2   Waterborne Diseases Caused by Flooding Disasters

Waterborne infectious diseases, particularly diarrhea, are a serious problem to people suffering from flood disasters. Dirty floodwater contaminated with pathogenic microorganisms is a key factor causing people to become infected. To prevent waterborne outbreaks, numerous studies have attempted to detect and assess the risk of waterborne diseases based on the direct measurement of pathogens, but the complicated and time-consuming laboratory testing causes a lack of samples for comprehensively modeling and analyzing the risk in flood-affected areas. In contrast, based on the back-propagation neural network (BPNN) technique, this study provides a new approach to assess the outbreak risk of diarrhea due to flooding by using simple parameters, including flood duration, dissolved oxygen (DO), and population density. Besides these results, the study also reveals the association between flood disasters and waterborne diseases based on multi-temporal remote sensing imageries.

However, since our prediction model relies on a BPNN algorithm, it still has some limitations in accordance with BPNN disadvantages. Although BPNN has many advantages to determine correlation between input and output data, BPNN has some drawbacks from the training model (Han and Kamber 2006). Some significant disadvantages of BPNN are also presented in the study. Firstly, BPNN has a slow rate of convergence(Witten and Frank 2005). Hundreds of thousands of times (epochs) are usually required for training the BPNN to converge at a specific accuracy. In the study, a large number of spatial input points (pixels) were input to train the BPNN model, thus it took a long time to meet a convergence for tuning parameters and defining an appropriated number of hidden layers. Secondly, BPNN sometimes provide local minima (Maimon and Rokach 2010). There are many local minima in the cost function curve (MSE and weights) of the BPNN. These local minima may reduce the BPNN's convergence rate or even prevent it from reaching the global minimum. Therefore, we need further studies to increase efficacy of the proposed method, such as developing an algorithm to level up the BPNN approach or comparing the BPNN with other machine learning approaches.

The main results of this study can be summed up following these points:

(1) The spatial distribution of dissolved oxygen (DO) based on IDW approach shows that the poorest quality of floodwater appeared in major industrial estates and some rural areas, while a better quality of floodwater was largely located in the countryside or in agricultural areas. The cross-validation error of IDW interpolation being in the range of acceptable assessment indicates that this method is appropriate to interpolate the spatial distribution of flood-related parameters.

(2) This study reveals the very close association between the temporal flood areas and the outbreak risk of waterborne diseases in the flood disaster. There are outbreaks of diarrhea, conjunctivitis and leptospirosis during the flood occurrence. Compared with the temporal in undation areas, the coefficient of determination (R^2) of the diarrheal risk ratios and the conjunctivitis risk ratio are 0.77 and 0.76 respectively.

(3) Modeling diarrheal outbreak risk in this study, the BPNN produced with very good prediction accuracy and high correlation coefficients, which measure the statistical correlation between predicted and reference RR, ranging from 0.79 to 0.93 (the correlation coefficient ranges from 1 for perfectly correlated results to 0 when there is no correlation) and acceptable RMS errors, ranging from 0.014 to 0.065. This indicates that our predictive models of the diarrheal-outbreak risk for each district are very accurate.

10.2 Further Work

Future works can be summed up into the following points:

(1) Implementing the SVM-PF method in other applications, for instance in urban area or forestry class separation. These could confirm that the SVM-PF method has a robust repertoire of performance uses.

(2) Another study may also be required to prove the robustness of the SVM-PF method, which is adding other variables to classification, or to use more dimensions in classification.

(3) While applying the state estimation technique in some other classification algorithms, in this approach, the investigation of the affecting parameter is needed before hand. However, a result of the improved method should present with better performance.

(4) Analyzing correlation between the temporal change of water quality data and the waterborne diseases that affect the prediction model can improve the efficacy of the prediction model. Because of the limitation on temporal resolution of the water quality data in the study, we need further study to detect its changes and effects to the infection of waterborne diseases.

(5) Comparing the BPNN model with other machine learning approaches, such as support vector machine- (SVM), Bayesian networks, and decision tree learning, can lead to increase the effectiveness of the proposed BPNN approach. Moreover, this may express the correlation between inputs and outputs more obviously leading to model the risk of waterborne diseases due to flooding more precisely.

(6) Other environmental factors affecting epidemics in a flood situation, such as immigration, the conditions of hygiene in various localities, and the flood intensity can be added to the model. Theoretically, adding more impact factors should produce better results, but on the other hand, this also would increase the required effort, cost, and time. The advantages and disadvantages must be weighed appropriately.

(7) Applying this approach to other flood related communicable diseases and/or other regions allows the possibility to be able to further evaluate the correlation of the flood related input parameters and the morbidity/mortality data. Furthermore, developing flood classified algorithms by integrating several techniques of multi-sensor and multi-temporal remote sensing imagery can lead to retrieve flood parameters more precisely and comprehensively.

References

Han J, Kamber M (2006) Data mining: concepts and techniques. Elsevier, San Francisco

Maimon O, Rokach L (2010) Data Mining and Knowledge Discovery Handbook. Springer, New York

Witten IH, Frank E (2005) Data mining: practical machine learning tools and techniques. Elsevier, San Francisco

Printed in the United States
by Baker & Taylor Publisher Services